Fundamentals of
Ground Engineering

Fundamentals of
Ground Engineering

John Atkinson

CRC Press
Taylor & Francis Group
Boca Raton   London   New York

CRC Press is an imprint of the
Taylor & Francis Group, an **informa** business

A SPON PRESS BOOK

CRC Press
Taylor & Francis Group
6000 Broken Sound Parkway NW, Suite 300
Boca Raton, FL 33487-2742

© 2014 by Taylor & Francis Group, LLC
CRC Press is an imprint of Taylor & Francis Group, an Informa business

No claim to original U.S. Government works

Printed on acid-free paper
Version Date: 20140411

International Standard Book Number-13: 978-1-4822-0617-3 (Paperback)

Library of Congress Cataloging-in-Publication Data

Atkinson, John, 1942-
    Fundamentals of ground engineering / author, John Atkinson.
        pages cm
    Includes bibliographical references and index.
    ISBN 978-1-4822-0617-3 (paperback)
    1. Earthwork. 2. Soil mechanics. I. Title.

TA715.A85 2014
624.1'5--dc23
                                           2014001335

Visit the Taylor & Francis Web site at
http://www.taylorandfrancis.com

and the CRC Press Web site at
http://www.crcpress.com

Contents

Contents

Contents

Geology and Engineering

Ground engineering combines geology and engineering. Normally ground engineers have first degrees in geology or in civil engineering and many have a post-graduate degree in engineering geology or geotechnical engineering.

During their first degrees, geologists are trained to observe the ground and the fossils and structures in it and draw inferences about the history of planet Earth. They make use of chemistry, botany, zoology and so on. They relate their observations to engineering performance largely through empirical correlations and transforms. During their first degrees, civil engineers are trained to make measurements of strength and stiffness of materials, steel, concrete and soil, and to predict the performance of structures. They make use of physics, mechanics and mathematics and use theories such as elasticity and plasticity to represent ground behaviour.

Each discipline has a complementary role to play in ground engineering. Geotechnical engineers should know enough geology to be able to communicate with geologists; geologists should know enough physics and mathematics to be able to communicate with geotechnical engineers.

There are several books on engineering geology that give largely empirical relationships between the geological history of the ground and its engineering behaviour and the behaviour of structures and excavations. This book describes the physical theories for the strength and stiffness of the ground and the analyses that are routinely used by engineers to predict the behaviour of slopes, foundations, and other structures in the ground.

These basic theories apply strictly to the behaviour of materials made of unbonded grains, which is a very good approximation for soil. The ground is taken to be elastic or plastic and frictional or cohesive and analyses satisfy the basic requirements of equilibrium and compatibility. Although natural soils may behave a little differently, their behaviour still closely follows the theories in this book. The outcomes should be designs that are safe, serviceable, economic, sustainable and buildable and the fundamental analyses make use of the basic theories in this book.

How to Use This Book

This book is not meant to be read from the first page to the last like a novel. The best way to use it is to select a topic for study and find what it says here. Then go to a book, or books, that deal with the topic in more detail but make sure that what you find there agrees with the simple basic theories of mechanics and the analyses that are in this book. Study is like navigating a city road map; there are many routes to the destination but at each turn you should know exactly where you are.

The book is not a collection of lecture notes although it covers the core geotechnical engineering content of typical undergraduate courses in civil engineering and post-graduate courses in geotechnical engineering and engineering geology. It is not a design guide although it includes some charts and analyses that are useful for design of geotechnical works. It is a summary, with few words but plenty of diagrams, of the fundamental theories and analyses that underpin all geotechnical engineering.

Things to Do

Geotechnical engineers should be familiar with fundamental principles and theories—most of them are relatively simple applications of basic mechanics to the behaviour of granular materials. But it is not enough to be able to do the mathematics; geotechnical engineers need to have an understanding of geology, knowledge of soil behaviour and a feel for what are reasonable numbers.

The photographs have been chosen to illustrate the general features of routine geotechnical engineering practice and they do not necessarily show state-of-the-art field and laboratory investigations and geotechnical works. Readers should search for other pictures or, better still, see the real thing.

At the end of most chapters there are suggestions for things to do to illustrate the principles in the chapter. Most do not require special equipment and can be done at home using kitchen equipment and samples commonly found in the home and garden. Schools once taught science students how to make simple test equipment and do simple experiments and that was part of a physics course. It is a good skill for an engineer to have.

Further Reading

This book should be used together with other books that students and engineers may already have or which their teachers and mentors recommend but deliberately there are very few specific references.

There are many long and expensive books on engineering geology, soil mechanics and geotechnical engineering. Some focus on theories and others on applications; many are at best unclear. They give advice that can appear to be conflicting or even wrong and it is difficult to recommend any one book or collection of books for further study.

Whatever your favourite book may say and whatever your teacher or mentor may say it ought to have a basis in physics and mechanics. Geotechnical engineering is a practical science with a theoretical basis; it is not a creative art and certainly not magic.

Acknowledgements

I have long thought that there was a need for a short and easily read book that set out the basic theories of geotechnical engineering. This book is influenced by Tony Waltham's *Foundations of Engineering Geology* but, because of the nature of the subject, it has a somewhat different style and approach.

Tony Moore, my editor, at Taylor & Francis has chased me for this book for several years and I am grateful to him for his persistence. Thanks are due to friends and colleagues who read and commented on drafts and to the printers and publishers who pieced together my drafts into the one page per topic format.

Many of the photographs are my own; I am grateful to Tony Waltham, David Norbury, Marcus Matthews, Richard Levine, Chris Eccles (TerraConsult), Bill Grose (Arup Geotechnics), George Tuckwell (RSK), Darren Ward (In Situ), Bill Howard (Griffin Soils Group) and Andrew Smith, Alex Booer and Helen Chow (Coffey Geotechnics) for additional photographs.

Units

In the SI system used in this book the basic units of measurement are

Length   m

Time   s

Force   N   multiples   kiloNewton $1\ kN = 10^3\ N$

megaNewton $1\ MN = 10^6\ N$

Some useful derived units are

Velocity   m/s

Acceleration m/s^2

Stress (pressure) $kN/m^2 = kiloPascal = kPa$

Unit weight kN/m^3

Unit force (1 N) gives unit mass (1 kg) unit acceleration (1 m/s²). The acceleration due to the Earth's gravity is $g = 9.81\ m/s^2$; hence, the force due to a mass of 1 kg at rest on Earth is 9.81 N. (*Note*: there are about 10 apples in 1 kg so a stationary apple applies a force of about 1 N acting vertically downwards.)

Symbols

(The symbols in this book may differ from those in other books and technical papers.)

σ	total normal stress
u	pore pressure
σ'	effective normal stress
$\tau = \tau'$	total and effective shear stress

In some books effective normal stress is denoted as $\bar{\sigma}$. In most books there is no distinction between total and effective shear stress and τ' is not used.

One dimensional and shear tests:

τ'	shear stress
σ'	normal stress
γ	shear strain
ε_v	volumetric strain = normal strain

Axisymmetric and triaxial tests:

$q' = (\sigma'_a - \sigma'_r)$	deviatoric stress
$p' = \frac{1}{3}(\sigma'_a + 2\sigma'_r)$	mean normal stress
$\varepsilon_s = \frac{2}{3}(\varepsilon_a - \varepsilon_r)$	shear strain
$\varepsilon_v = \varepsilon_a + 2\varepsilon_r$	volumetric strain

Plane strain ($\varepsilon = 0$ out of the page):

$t' = \frac{1}{2}(\sigma'_v - \sigma'_h)$;

$s' = \frac{1}{2}(\sigma'_v + \sigma'_h)$

$\varepsilon_\gamma = (\varepsilon_v - \varepsilon_h)$;

$\varepsilon_{vol} = (\varepsilon_v + \varepsilon_h)$

Superscripts for strains:

e	elastic
p	plastic
c	creep

Subscripts for states:

0	initial state (i.e. τ'_0)
c	critical state (i.e. τ'_c)
p	peak state (i.e. τ'_p)
y	yield stress (i.e. σ'_y)
m	limiting stress ratio (i.e. $\tan\phi'_m$)
r	residual (i.e. $\tan\phi'_r$)

Subscripts for axes:

v, h	vertical and horizontal
a, r	axial and radial

A	area; activity; parameter for peak state power-law envelope
B	breadth or width
C	compliance
C_c	slope of the normal compression line
C_α	creep parameter
C_s	slope of a swelling and recompression line
D	depth
D_r	relative density
D_w	depth of water
E	work done by external loads
E	Young's modulus (E′ for effective stress; E_u for undrained loading)
F	load on a foundation
F_a	axial force
F_n	normal force
F_s	shear force
F_s	factor of safety
G	shear modulus (G′ for effective stress; G_u for undrained loading)
G_o	shear modulus at very small strain

G_s	specific gravity of soil grains
H_c	critical height (of a slope)
H	height or thickness; maximum drainage path; hardening parameter $\delta\tau'_y/\delta\gamma^p$
I_L	liquidity index
I_P	plasticity index
I_σ	influence coefficient for stress
I_ρ	influence coefficient for settlement
J	stiffness modulus that couples shear and volumetric parameters
K'	bulk modulus
K_0	coefficient of earth pressure at rest
K_a	coefficient of active earth pressure
K_p	coefficient of passive earth pressure
L	length
L_f	load factor
M'	one-dimensional modulus
N_s	stability number (for undrained slopes)
N_f	number of flow channels (in a flownet)
N_d	number of equipotential drops (in a flownet)
N	normal force
N_c, N_γ, N_q	bearing capacity factors
P	potential $= z + u/\gamma_w$
P_a	force due to active pressure
P_p	force due to passive pressure
P_w	force due to free water pressure
Q	flow (volume); pile load
Q_b	pile base resistance
Q_s	pile shaft resistance
R	radius
R_o	overconsolidation ratio $= \sigma'_m/\sigma'$
R_y	yield stress ratio $= \sigma'_y/\sigma'$
S	stiffness: degree of saturation
S_σ	state parameter $= \sigma'/\sigma'_c$
S_w	state parameter $= w_\lambda - w_\Gamma$

T	shear force; surface tension force
T_v	$= c_v t/H^2 =$ time factor for one-dimensional consolidation
U	force due to pore pressures
U_t	average degree of consolidation after time t
V	volume; velocity (of seepage)
V_w	volume of water
V_s	volume of soil grains
W	weight; work dissipated by interval stresses
W_w	weight of water
W_s	weight of soil grains
a	acceleration
b	thickness or width: parameter for peak state power-law envelope
c'	cohesion intercept in Mohr–Coulomb failure criterion
c_v	coefficient of consolidation for one-dimensional consolidation
e	voids ratio; eccentricity
e_0	voids ratio of normally consolidated soil at $\sigma' = 1.0$ kPa
e_κ	voids ratio of overconsolidated soil at $\sigma' = 1.0$ kPa
e_Γ	voids ratio of soil on the critical state line at $\sigma' = 1.0$ kPa
h_w	height of water in standpipe
i	slope angle; hydraulic gradient
i_c	critical slope angle; critical hydraulic gradient
k	coefficient of permeability
m	mass
m_v	coefficient of compressibility for one-dimensional compression
m, n	slope stability numbers (for drained slopes)
p'_m	maximum past stress
p_w	free water pressure
q	bearing pressure; rate of seepage
q_c	bearing capacity
q_n	net bearing pressure
q_a	allowable bearing pressure
r	radius

r_u	pore pressure coefficient $= u/\sigma'_v$
s	length along a flowline
s_u	undrained strength
t	time
u	pore pressure
u_a	pore air pressure
u_w	pore water pressure
u_∞	long-term steady-state pore pressure
u_0	initial steady-state pore pressure
$\bar{u}$	excess pore pressure
v	specific volume
v_κ	specific volume of overconsolidated soil at $p' = 1.0$ kPa
w	water content
w_L	liquid limit
w_P	plastic limit
w_λ	$w + C_c \log \sigma'$
w_Γ	water content of soil on the critical state line at $p' = 1.0$ kPa
Γ	specific volume of soil on the critical state line at $p' = 1.0$ kPa
Δ	large increment of ....
M	slope of CSL projected to $q'{:}p'$ plane
N	specific volume of normally consolidated soil at $p' = 1.0$ kPa
Σ	sum of ....
α	factor for undrained shear stress on pile shaft
γ	unit weight
γ_d	dry unit weight
γ_w	unit weight of water ($= 9.81$ kN/m^3)
δ	small increment of ....
δ'	angle of friction between structure and soil
η	stress ratio q'/p'
κ	slope of swelling and recompression line
λ	slope of normal compression and critical state lines
ν	Poisson's ratio (ν' for drained loading, $\nu_u = 1/2$ for undrained loading)

ρ	settlement
ρ_c	consolidation settlement
ρ_i	initial settlement
ρ_t	settlement at time t
ρ_∞	final consolidation settlement
ρ_a	allowable settlement
ρ	density
ρ_d	dry density
ϕ'	angle of friction
ϕ'_c	critical state angle of friction
ϕ'_p	peak angle of friction
ϕ'_m	mobilised angle of friction
ϕ'_r	residual angle of friction
ψ	angle of dilation
τ, τ'	total and effective shear stress (usually with a subscript)
σ, σ'	total and effective normal stress (usually with a subscript)

Chapter One

What is Geotechnical Engineering?

Geotechnical engineering, with structural engineering and hydraulic engineering, is one of the major design components of civil engineering. Structural engineering requires structural analysis and knowledge of man-made materials. Geotechnical engineering requires geotechnical analysis and knowledge of natural soils, rocks and groundwater. The fundamental principles for design of groundworks use many of the basic theories and methods familiar in other branches of engineering.

1.1 CONSTRUCTION IN THE GROUND

Geotechnical engineering is construction on and in the ground using natural soils and rocks. Foundations add load to the ground and settle; excavations remove load and the ground heaves. Man-made slopes should remain stable but may need support from retaining walls. Dam slopes must remain stable and the dam must retain water. Road pavements are made from soil and rock.

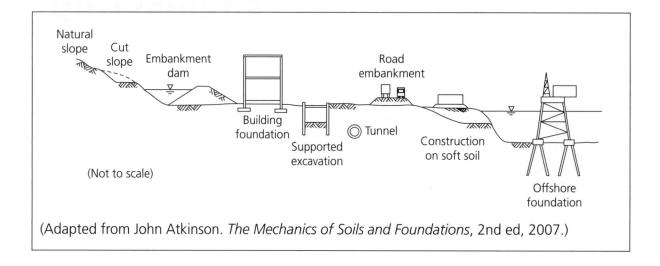

(Not to scale)

(Adapted from John Atkinson. *The Mechanics of Soils and Foundations*, 2nd ed, 2007.)

1.1 CONSTRUCTION IN THE GROUND (CONT)

Shallow foundations

Stable slopes

Retaining walls

Deep foundations

Dams

Pavements

All these constructions require geological investigations to discover what the ground is made of, testing to determine its strength and stiffness and analyses of stability, movement and seepage of water.

1.2 WHO ARE GEOTECHNICAL PROFESSIONALS AND WHAT DO THEY DO?

Geotechnical professionals usually start as geologists, engineers or scientists and specialise as they progress through their education and training. Each brings different skills in investigation, analysis and design.

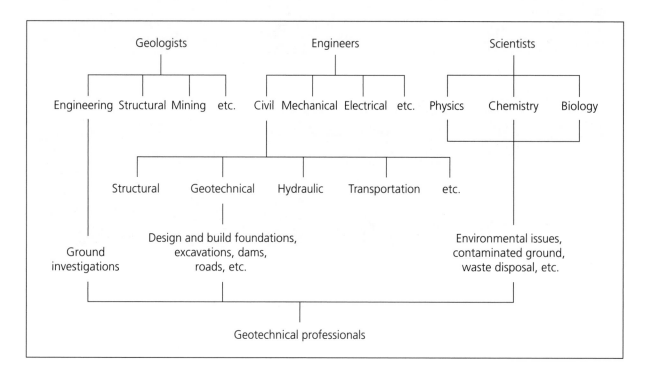

Principal activities of		
Engineering geologists	**Geotechnical engineers**	**Scientists**
Investigate the ground: observe and map surface features; excavate and observe boreholes and test pits.	Test soils in the ground and in the laboratory.	Investigate and clean up contaminated ground.
Assess properties and behaviour from observation and description.	Assess properties and behaviour from test results.	Design waste repositories.
Create a geological model.	Create a geotechnical model.	Assess chemical and biological interactions between buried structures and the ground.
Assess sources of construction materials.	Analyse and design foundations and walls and assess stability of slopes.	Use chemicals to modify and improve ground properties.

1.3 EDUCATION AND TRAINING

Geotechnical professionals acquire their skills through a combination of undergraduate and post-graduate study and on-the-job training, mentoring and experience. Most have first degrees in civil engineering or geology and occasionally science. Many geotechnical professionals have a post-graduate degree in a relevant subject such as engineering geology or geotechnical engineering.

	Engineering geologist	Geotechnical engineer	Scientist
Undergraduate	Degree in geology Chemistry Mineralogy Observation and mapping Trained to observe and record	Degree in civil engineering Maths Physics Mechanics Trained to analyse and design	Degree in one or more of Maths Physics Chemistry Biology
Post-graduate	Engineering geology Principles of geology applied in a civil engineering context Geological model; identify principal strata; locate strata boundaries, identify and locate structural features (folds, faults); identify groundwater regime Locate and assess materials and aggregate resources; empirical design of groundworks	Geotechnical engineering Principles of mechanics applied to soils, rocks and groundwater Geotechnical model; identify ground types and groundwater pressures, assign engineering parameters Analyse and design groundworks; foundations, slopes and walls, tunnels, pavements, etc.	Environmental sciences Geophysics

1.4 WHAT GOES WRONG?

Geotechnical structures sometimes fail, often because a geotechnical professional made a mistake. Sound understanding of basic principles should ensure this does not happen.

Structure	Example	Comment
Foundations settle too much		Grain elevator settled and tilted shortly after filling.
Slopes slip		Landslide in glacial soil caused by coastal erosion and groundwater pressures.
Walls collapse		Collapse of strutted retaining walls supporting a deep excavation in soft clay.
Excavations flood		Excavation in sands and gravels flooded by seepage from groundwater.

1.4 WHAT GOES WRONG? (CONT)

Structure	Example	Comment
Dams leak		Concrete lining leaked and seepage eroded a channel through the earth-fill dam.
Tunnels collapse		Collapse of tunnel during construction.

1.5 THINGS TO DO

On your next journey, which may be by road or rail or on foot, observe and make a list of all the structures that involve natural soil and rock for support or for a source of materials.

Bridges and buildings have foundations; roads and rail have excavated slopes and built embankments some of which may be supported by walls; road pavements are mostly compacted soil; railway ballast is crushed rock. You may also see dams and river banks retaining water.

Search the Web for 'geotechnical failures' or something similar and find examples. You may be able to find some of the cases shown in Section 1.4. Discover why these failures occurred.

1.6 FURTHER READING

There is a rich history of civil engineering and geotechnical engineering. Some of the founding fathers of geotechnical engineering include Coulomb, Rankine, Terzaghi, Skempton, Bishop, Peck, D.W. Taylor, Schofield and Wroth. Find out about these people and what they did.

Part A

Discovering the Ground

Chapter Two

Formation of Soils and Rocks

The ground was formed by the natural geological processes of weathering, erosion, transportation and deposition all influenced by plate tectonics. Geotechnical engineers should have a basic understanding of the geological processes that lead to formation of soils and rocks.

Geotechnical engineers do not need to be able to identify specific rocks but they need to be able to communicate with engineering geologists and understand how the formation and characteristics of soils and rocks determines their engineering properties.

2.1 BASIC SOIL AND ROCK CYCLE

Rocks near the surface weather and become residual soils. These are eroded, transported, re-deposited and buried under successive sediments. With increased pressure and temperature they become rocks again so there is a continuous cycle between rocks and soils. This is the principle of the geological cycle.

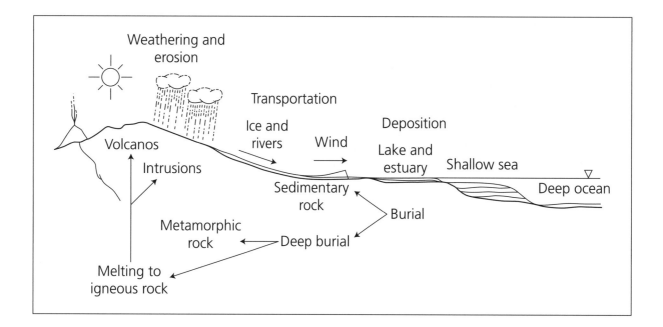

Plate Tectonics

The Earth is a dynamic place where everything is on the move. Plates of crust move at about 1 cm per year (i.e. 10 km in 1 million years). Plates collide forming mountains and separate allowing hot molten rock to come to the surface. These form igneous rocks as they cool and metamorphic rocks as they intrude into and heat nearby rocks.

Weathering, Transport, Deposition, Burial

Under attack by frost, sun and rain, surface rocks break up and form soils. These are transported by gravity, rivers, glaciers and wind and deposited as sediments. Soils are deposited at rates of the order of 1 mm per year (i.e. 1 km of depth in 1 million years). Stresses due to deep burial compress soils and they become sedimentary rocks.

2.2 GEOLOGICAL AGE

The fundamental classification of rocks used by geologists is based on age and this is world-wide. For example, a rock classified as Triassic is in the Mesozoic era of Earth's history and would have been deposited a little less than 250 million years ago. In the United Kingdom it might be part of the Mercia Mudstone Group and it can be a mudstone, a siltstone or a sandstone. Elsewhere in the world a rock of Triassic age would probably be completely different.

Age (My)	Era	Period	Examples in UK
	Quaternary	Recent Pleistocene (ice ages)	Fens Boulder clays and tills
2	Tertiary	Pliocene to Paleocene	East Anglian Crags London Clay
65	Mesozoic	Cretaceous Jurassic Triassic	Chalk Portland stone Mercia Mudstone
250	Paleozoic	Permian Carboniferous Devonian Silurian and Ordovician	Sandstones Coal measures Old Red Sandstone Slates in Wales
600	Precambrian		North-west Scotland

Geological age, on its own, does not determine the type of rock or soil, it simply defines when the ground was deposited or formed by an igneous process.

Soil and rock type may be linked to geological age at a specific location or within a relatively small region but even then soil and rock type can vary significantly.

For ground engineering purposes soils and rocks are best described and classified objectively using the schemes in Chapter 3.

2.3 NATURAL EARTH MATERIALS

Most geotechnical engineering is in the top few 10s of meters of the ground. The materials are soils and rocks and the works are usually within the influence of groundwater. Soils and rocks consist of the same basic materials. Rocks are relatively stiff and strong; soils are relatively soft and weak (see also Section 3.1).

	Engineering soils	**Rocks**
	Glacially deposited soils with coarse and fine grains	Sedimentary rock with bedding and faulting.
Classes	Fine-grained silt and clay Coarse-grained sand and gravel Well-graded soil: mixed coarse and fine	Igneous Sedimentary Metamorphic
Origins	Deposited from: • Water; alluvial • Wind; aeolian • Ice; glacial	Formed by: • Cooling of molten rock • Heating and pressure • Deep burial of soils
Characteristics	Individual grains unbonded or only very weakly bonded May be deposited in layers of different soils	Mineral crystals and very strongly bonded soil grains Large blocks separated by joints and fissures
Engineering properties	Relatively weak and soft • Strength < 100 kPa • Stiffness < 10 MPa Acts as a continuum Clay is nearly 'watertight' Sand and gravel is 'free draining'	Relatively strong and stiff • Strength > 1 MPa • Stiffness > 10 MPa Behaviour controlled by joints

2.4 SURFACE PROCESSES

Soils that are deposited depend significantly on the local climate at the time. This influences the weathering and transportation processes.

The main climatic regions on Earth are glacial and periglacial (near the poles and at high altitude); temperate; arid (low rainfall); tropical (hot and wet).

Climate		Weathering	Erosion and transport	Soils deposited
Glacial and periglacial		Expansion and contraction by freezing and thawing produces mostly silt, sand and gravel	By moving glaciers and by rivers in periglacial regions	Well-graded boulder clays and tills. Sands, gravels, silts and clays from periglacial rivers
Temperate		Little happens	By rivers	Sands and gravels from fast flowing rivers. Silts and clays from slow flowing rivers
Desert		Expansion and contraction by day-time heating and night-time cooling produces mostly silt, sand and gravel	By wind and occasional flash flooding	Wind-blown sands in dunes. Salts from evaporation
Tropical		Hot and wet conditions cause chemical changes and production of clay minerals	By rain and rivers	Mostly silts and clays

Earth's climate has changed several times over geological time from hotter to colder than at present so surface processes have changed too. In addition plate movements take crust into different climatic zones.

Any one region of Earth's surface today has probably experienced most of the possible climatic zones in its past and the ground will consist of many different soil and rock types.

2.5 LOCAL CLIMATE AND ROCKS IN ENGLAND

Most of the fragment of crust that we today call England has moved steadily northwards from a position south of the equator 500 my ago to its present position at about 60° N. It has passed through several climatic zones within each of which characteristic soils and rocks were formed.

Period (See Section 2.2)	Location of UK crust in present day	Some examples in England
Recent to Paleocene	Western Europe	Fens Boulder clays and tills London clay
Cretaceous and Jurassic	Mediterranean; warm shallow sea	Chalk, Portland Stone
Triassic and Permian	North Africa and Sahara; sandstones, mudstones and evaporates	Mercia Mudstone, Sherwood Sandstone
Carboniferous	Central Africa rainforest	Coal measures, Pennines
Devonian	South African deserts	Old Red Sandstone, Cornwall, parts of Devon, Brecon Beacons

Similar histories of crustal movement and local climate can be created for any location on Earth today and the history of the formation of the soils and rocks related to this prehistoric record.

2.6 THINGS TO DO

Wherever you are now, at home, in your office, or university, or on site, obtain a geological map and related literature on the local geology.

1. Imagine you are drilling a deep borehole. Draw a log for your borehole. On it mark the soils and rocks through which the borehole passes and give as much information about them as you can find.

2. Your location is within a region of crustal plate. Trace the movements of your location around Earth over geological time and so relate the local climate to geological age.

3. Relate the soils and rocks found in your borehole to the local climate at the time they were deposited.

When you are investigating the ground you should establish (a) where in the world materials like the ones below you are being deposited today and (b) what has happened to them since they were deposited.

2.7 FURTHER READING

These are many books on geology and engineering geology. The best ones are those that give simple and clear coverage of basic principles with plenty of pictures, diagrams and maps.

Chapter Three

Description and Classification of Soils

Before any groundworks are designed the ground has to be investigated and its behaviour and properties determined. This is different from structural engineering where the materials, usually concrete and steel, are specified by the designer and manufactured to have the required properties.

The first stage of a ground investigation (Chapter 4) is to describe the ground as seen in exposures, test pits and samples and from very simple tests. Many of the words used to describe soil have specific meanings and it is important that these are used consistently and correctly.

Some basic soil properties of strength and stiffness can be derived from these simple descriptions (Section 14.6).

3.1 DIFFERENCES BETWEEN SOILS AND ROCKS

There are basic differences between the properties and behaviours of soils and rocks that are important for geotechnical engineering and for soil and rock mechanics (see also Section 2.3).

Soils	Rocks
Grains are unbonded or only weakly bonded. When immersed in water grains disaggregate or the soil becomes very soft.	Grains and crystals are very strongly bonded.
Structural features such as fissures and bedding do not greatly influence strength and stiffness of the soil mass.	Joints have a significant influence on the strength and stiffness of the rock mass.
Slip surfaces do not exist before failure but are generated by the failure.	Failures occur on joints and discontinuities that existed before the failure.

3.2 BASIC DESCRIPTION OF SOILS AND ROCKS

Soils and rocks are described from samples and from exposures where the ground can be seen in natural or excavated slopes and pits. Descriptions are mostly qualitative and are made from visual observation and very simple tests. They lead to approximate estimates of engineering properties of strength, stiffness and permeability (Section 14.6).

Feature	Soil	Rock
What it is	The grains: • Grading and sizes • Shape and surface texture • Mineralogy	Geological names: • Age (e.g. carboniferous) • Origin (e.g. igneous) • Type (e.g. granite)
State	How the grains are packed: • Dense or loose	Weathering: • Fresh rock to residual soil
Structure	Bonding: • Unbonded or weakly bonded Fabric: • Homogeneous or layered	Joints and faults: • Orientation • Spacing • Roughness

3.3 GRADING OF SOIL

Grading is the distribution of the grain sizes and is the most important feature of a soil. Grain sizes are given names:

Clay	Silt			Sand			Gravel		
	Fine	Medium	Coarse	Fine	Medium	Coarse	Fine	Medium	Coarse

0.002 0.006   0.02   0.06   0.2   0.6   2   6   20   60 mm

Fine sand (>0.06 mm) can be seen with the naked eye but coarse silt cannot. The word 'clay' has two meanings. Here, clay means a grain smaller than 0.002 mm whatever it is made of; the word clay can also mean a clay mineral such as kaolinite which can have grains larger than 0.002 mm.

Distributions of particle sizes are described by grading curves. The vertical axis is the per cent by weight of grains less than the given size.

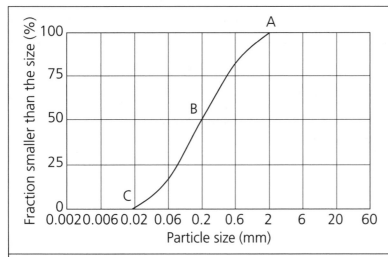

At point A, 100% is smaller than 2 mm (so 2 mm is the largest size).

At point B, 50% is smaller than 0.2 mm.

At point C, nothing is smaller than 0.02 mm (so 0.02 mm is the smallest size).

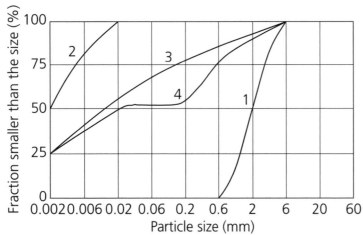

Soil 1 is poorly graded sand and gravel.

Soil 2 is poorly graded silt and clay.

Soil 3 is well graded.

Soil 4 is gap graded; there is a gap in the grading and sizes from 0.02 to 0.2 mm are missing.

3.3 GRADING OF SOIL (CONT)

There are simple procedures to determine grading. The soil must first be disaggregated and, for grading by eye and by sieving, it must be dried.

By Eye	Sedimentation	Sieving
Separate the grains into heaps of visible and invisible sizes.	Shake the cylinder (which is full of soil and water) horizontally so the grains are distributed.	The sieves have apertures of decreasing size down to about the silt–sand division.
Determine the maximum size.	Place the cylinder upright and allow the grains to sediment.	Place the sample on the top (largest) sieve and shake.
Estimate the proportions of the heaps.	Observe the thicknesses of the layers of different sizes.	Weigh the fraction retained on each sieve.
Plot the maximum size and the fractions at 0.06 mm (smallest visible size).	Grains that remain in suspension after a few hours are clay sized.	Determine fractions of coarse, medium and fine silt and clay by analytical sedimentation.

Soil grading is often described in words such as silty clay or sandy gravel. When describing soil grading it is best to sketch a grading curve by eye and when you see a word description you should sketch a grading curve.

3.4 DESCRIPTION OF SOIL BASED ON GRADING

Grading determines how water behaves in soil particularly drainage and suction (Section 7). Soils should be classified as either coarse grained or fine grained.

Coarse-grained soil: less than about 30% fines	Fine-grained soil: more than about 40% fines
There are not enough fine grains to completely fill the spaces between the coarse grains.	The fine grains completely fill the spaces between the coarse grains.
Water can travel quickly through large pore spaces.	Water can only travel slowly through small pore spaces.
Soil can only sustain relatively small suctions (Section 7.3).	Soil can sustain relatively large suctions.

Warning: Soils are often described as cohesive or granular or frictional; this is not helpful. They should be described as either fine grained or coarse grained because this controls drainage which has a major influence on soil behaviour (Section 8.4).

3.5 CHARACTERISTICS OF COARSE SOIL GRAINS

Coarse grains are larger than about 0.1 mm and they can be seen with the naked eye.

	Observation	Inference
Shape and texture	Rounded and smooth	Deposited from water or wind Relatively small frictional strength
	Angular and rough	Deposited from glaciers or residual soil Relatively large frictional strength
Colour	Colourless or translucent	Quartz: hard and strong
	Pale colour	Calcium carbonate: brittle
	Brown or grey	Broken rock

3.6 CHARACTERISTICS OF FINE SOIL GRAINS

Fine grains (silt and clay sizes together) are too small to see with the naked eye. They are described by their behaviour and especially the range of water content between very high water content (w_L = liquid limit) and very low water content (w_P = plastic limit).

Liquid limit w_L Water content at which the soil flows like a liquid	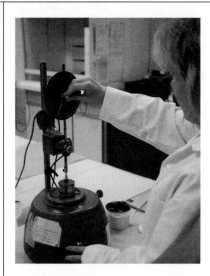	Determined by penetration of a weighted cone. Corresponds to a strength of around 2 kPa.
Plastic limit w_P Water content at which the soil becomes brittle and crumbles		Determined by rolling a thread. Corresponds to a strength of around 200 kPa.

Plasticity index $I_P = w_L - w_P$. This is the range of water content over which a soil behaves as a soil; it neither flows like a liquid nor becomes brittle and crumbly. This ranges from nearly zero (mostly silt sized) to over 100 (mostly active clay). I_P is related to compressibility and to friction angle (Section 14.6).

3.7 WATER CONTENT UNIT WEIGHT RELATIONSHIPS

Soils consist of grains with water and perhaps air in the pore spaces. In saturated soil there is no air, only water. The basic theories of soil mechanics and geotechnical engineering are for saturated soils. There are simple relationships linking the quantities of soil grains and water.

		Volumes	Weights	Measurements
Air		V_a	$W_a = 0$	Measurements are made on soil samples of volume V and weight W and, after drying, W_s
Water		V_w	$W_w = \gamma_w V_w$	
Grains		V_s	$W_s = G_s \gamma_w V_s$	The weight of water is found from $W_w = W - W_s$
Totals		V	W	

Name	Symbol	Relationship	Typical Values for Soil
Water content	w%	$\dfrac{W_w}{W_s}$	10% to >60%
Unit weight	γ kN/m³	$\dfrac{W}{V}$	18 kN/m³ to 22 kN/m³
Voids ratio	e	$\dfrac{V_a + V_w}{V_s}$	0.2 to >1
Specific volume	v	$\dfrac{V}{V_s} = 1 + e$	1.2 to >2
Degree of saturation	S%	$\dfrac{V_w}{V_a + V_w}$	0 = dry and 100% = saturated

Soil grains and water are incompressible so volume change in saturated soil is equal to the volume of water that drains out of or into the soil. Volumetric strain (compression is positive) is

$$-\delta\varepsilon_v = \frac{\delta V_w}{V} = \frac{\delta v}{v} = \frac{\delta e}{1 + e}$$

3.8 PACKING OF GRAINS

Soil grains may be packed together loosely or densely. Dense soil has a low water content and is relatively stiff and strong; loose soil has a high water content and is relatively soft and weak.

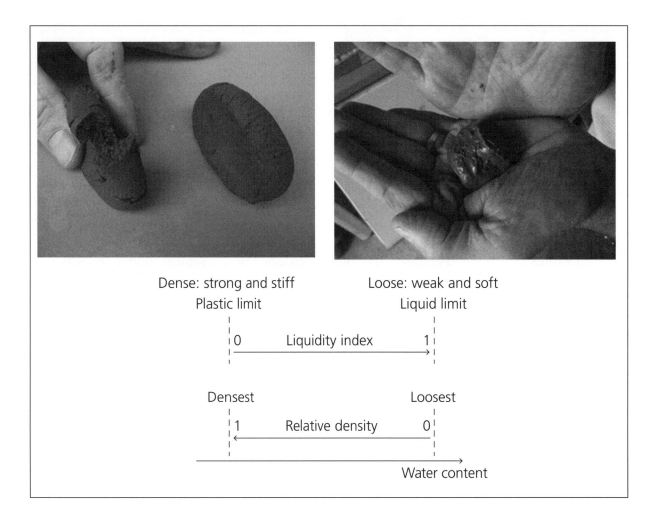

The state of packing is related to the maximum and minimum values of water content or voids ratio.

Liquidity index $I_L = \dfrac{w - w_P}{w_L - I_P}$ ranges from 1 (at the liquid limit) to 0 (at the plastic limit).

Relative density $D_r = \dfrac{e_{max} - e}{e_{max} - e_{min}}$ ranges from 0 (loosest state) to 1 (densest state).

3.8 PACKING OF GRAINS (CONT)

The liquidity index of a fine-grained soil is described by simple manipulation tests and these are related to its undrained strength (Section 11.4).

Description	Undrained strength (kPa)	Description	Liquidity index
Liquid limit	≈2	Flows	1.0
Very soft	<10	Extrudes between fingers when squeezed	
Soft	10–20	Can be moulded by light finger pressure	~0.5
Firm	20–40	Moulded by strong finger pressure	~0.4
Stiff	40–75	Cannot be moulded; can be indented by thumb	~0.3
Very stiff	75–150	Can be indented by thumbnail	~0.2
Hard	>150	Difficult to mark with thumbnail	<0.1
Plastic limit	≈200	Crumbles	0

These descriptions for fine-grained soils vary a little between different countries and regions but the basic principles and scales appear in most standards and codes.

3.9 STRUCTURE IN SOIL

Structure in soil is the combination of fabric and bonding; structure is the difference between soil in the ground and the same soil grains completely disturbed but at the same stress and water content.

Type	Example	Investigate	Effect on engineering behaviour
Fabric: (bedding and layering)		Observe in exposures or split samples.	Permeability very different for horizontal and vertical seepage.
Bonding and cementing		Immerse sample in water. Unbonded sand quickly forms a cone. Unbonded fine-grained soil slowly becomes very soft. Cemented and bonded soils maintain strength.	Additional strength and stiffness; but only for small strains. Deformation destroys bonding and soil is brittle.

Fabric is mostly due to non-uniform deposition resulting in layering of coarser and finer grains. Bonding is mostly due to deposition of salts at the points of contact between grains.

Not all natural soils have significant structure. In some cases structure can be removed by natural geological remoulding such as by moving ice sheets. To investigate structure split the sample along its length and examine the split surfaces with a hand lens.

3.10 THINGS TO DO

There are many opportunities for examining the topics in this section by simple experiments and observations. Complicated equipment is not required; all that is needed are kitchen scales, some jam jars, a ruler and an oven or microwave. (Drying soil in a microwave is not normally permitted in codes but it is a quick way to dry soil and get a reasonable value for water content.)

Get some samples of different soils from locations such as a garden, a beach, a river bank, a builder's yard, a soils laboratory. In the kitchen granular materials such as sugar and coffee beans are like coarse-grained soils while flour and ground coffee are like fine-grained soils.

1. For each soil estimate the grading curve by eye and by sedimentation as described in Section 3.3. A tall jam jar can be used instead of a measuring cylinder.

2. Take some coarse-grained soils, or the coarse fraction of well-graded soils, and describe the characteristics of the visible grains. How would you describe the grains of granulated sugar, caster sugar and icing sugar?

3. Take some fine-grained soils, or the fine fractions of well-graded soils. Dry the soil by gentle heating in an oven or by heat from your hand until it becomes crumbly. Add water until the soil becomes like a liquid and flows. At each state (crumbly and flowing) measure the water content by weighing and drying a sample in an oven or microwave.

4. Start with some fine-grained soil at a low water content and gradually add water and mould it in your hand. Observe how the soil becomes easier to mould as the water content increases and relate this to the table in Section 3.6.

5. Take a known weight of dry sand and fill a jam jar with a known diameter about 3/4 full. (The internal diameter of a jam jar can be measured in several different ways.) Shake it up and put it down gently; the 'soil' is now loose. Measure the height of soil and hence find its volume. Tap the jam jar gently so the soil compresses due to vibration; the soil is now dense. For each condition, dense and loose, determine the voids ratio, specific volume and unit weight. (For this exercise assume the specific gravity of the sand grains to be 2.7.)

3.11 FURTHER READING

The tests that are used in practice for soil description and classification are strictly determined by national codes and specifications. Discover which codes are current in your region and study their requirements for description and classification of soils.

The best geotechnical professionals have spent some time in a commercial or research soils laboratory describing soil samples.

Chapter Four

Ground Investigations

The ground is there as it is and it is necessary to discover what it is by ground investigations. These involve activities in the office, on site and in the laboratory and there is a clear process to be followed.

In many cases, the basic nature of the ground and approximate values for ground properties can be found from initial study of geological records and by using relationships between soil description and engineering behaviour. Work on site and in the laboratory should be designed to confirm or to disprove or to modify these initial findings.

The outputs of a ground investigation are reports. These contain idealisations (models) of the ground. They should contain sufficient information for safe and economic design of groundworks.

4.1 GROUND MODELS

A ground model is a 3D representation of the ground shown as plans and sections. It should show the significant soils and rocks and the groundwater.

The information in a ground model depends on its purpose and the two main ground models for geotechnical engineering are geological models and geotechnical models. Groundworks are designed on the basis of the geotechnical model.

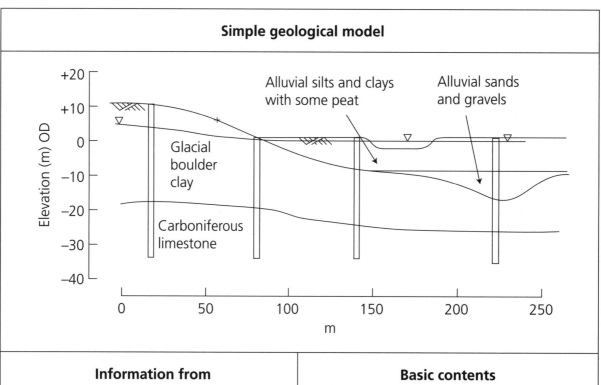

Simple geological model

Information from	Basic contents
Geological maps and memoirs	Geological formations (using geological names)
Site inspections	Geological descriptions of all geological formations
In situ testing	
Drilling and visual identification of samples	Locations of boundaries between geological materials
	Structure: folds, faults
	Geological history

4.1 GROUND MODELS (CONT)

Warning: It is important to distinguish between information that is *factual* (and comes from direct observation and measurement) and information that is *interpreted* from factual information.

For example, the geological profile between boreholes is interpreted from factual information that is normally obtained only from observations on samples from boreholes.

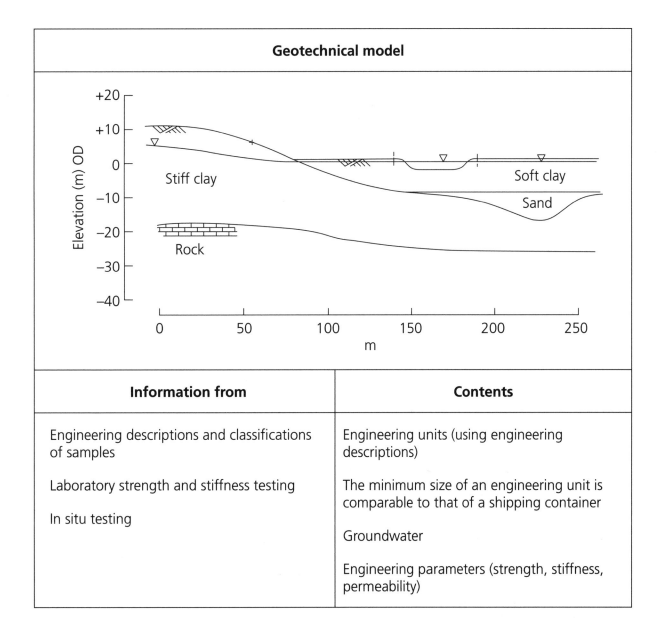

Geotechnical model

Information from	Contents
Engineering descriptions and classifications of samples	Engineering units (using engineering descriptions)
Laboratory strength and stiffness testing	The minimum size of an engineering unit is comparable to that of a shipping container
In situ testing	Groundwater
	Engineering parameters (strength, stiffness, permeability)

4.2 METHODOLOGY FOR GROUND INVESTIGATIONS

Ideally a ground investigation and development of a ground model is a process of evaluation of current information which then guides further work; in practice, the opportunities for feedback and further work are often limited.

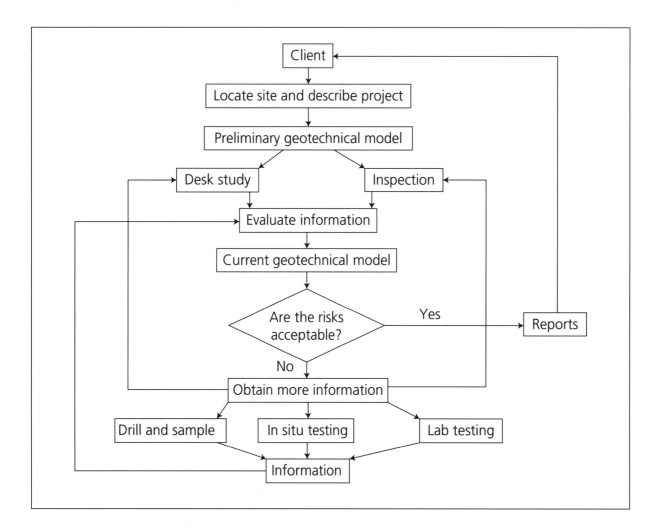

The preliminary geotechnical model is created on the basis of the client's brief and prior knowledge of the ground conditions before any investigations are done—this model is necessarily approximate. All subsequent work is aimed at refining and adding to the preliminary model.

It is most unlikely that any geotechnical model will be exactly correct; the question is the balance between risks from uncertainties and the additional costs of conservative design.

4.3 REPORTS

The outputs of ground investigations are reports; these are of three kinds depending on the information in them.

Report	Contents
Factual report	Records of: desk studies and walk-over inspections; borehole and test pit logs. Several borehole logs can be shown together on a cross section but lines should not be drawn between them because this requires interpretation.
	Objective descriptions of soil, rock and groundwater samples.
	Factual results of in situ and laboratory tests. These include grading curves, Atterberg limits, observations of manipulation of hand samples, undrained strength from vane tests.
	Factual results of triaxial, shear and 1D consolidation tests are stress and strains: derivation of strengths and stiffnesses from these data require interpretation.
Interpretive report	3D geological model as plans, sections and block diagrams showing the locations of geological formations and geological structures such as folds and faults.
	3D geotechnical model showing locations of ground with the same, or very similar engineering behaviour and properties.
	Values of soil parameters such as strength, stiffness and permeability interpreted from the test results given in the factual report. These may be plotted on cross sections or against depth or level
	Values of soil parameters are often included in the interpretive report.
Design report	The geotechnical model used in the design of the works.
	This includes the 3D geotechnical model developed from the interpretive report but includes values for the soil parameters used in the design calculations.

4.4 COMPONENTS OF GROUND INVESTIGATIONS

Activity	Examples
Desk study of existing records	
Mapping and inspection of exposures, test pits and samples	
Drilling and sampling	
In situ testing	
Laboratory testing	

Common methods	Outcomes
Topographical maps—recent and historical Geological maps and memoirs Literature searches Air photographs—recent and historical Records of nearby investigations and works	Preliminary ground model • Plans and sections • Estimates of behaviour and properties
Observe and map surface features Log exposures Samples from exposures and shallow pits	Refined preliminary ground model
Drilling in soft ground and rock Additional test pits Obtain samples: • Disturbed samples • Samples in driven and pushed tubes • Hand-cut blocks from exposures and test pits • Samples of groundwater	Factual borehole and test pit logs Samples for testing
Penetration probing • Standard penetration test SPT • Cone penetration test CPT Loading tests • Vane shear • Pressuremeter Geophysics • Seismic profiles and shear wave velocity	Ground properties from probing, loading and geophysics (Section 14.2)
Description and classification • Particle characteristics – grading and plasticity • State – relative density • Microscopy and chemical Strength and stiffness • Triaxial • Shear • 1D consolidation	Engineering descriptions of samples Factual data from triaxial, shear and 1D consolidation testing (Sections 9.1 and 14.2)

4.5 RELIABILITY OF THE GEOTECHNICAL MODEL

Geotechnical engineers designing groundworks should have an understanding of the reliability of the ground information on which they base their designs.

The geological model is built from observations of the ground in exposures, test pits and boreholes together with the geological history of the location. There are clearly uncertainties associated with extending information obtained at specific locations to other locations.

The geological model, with its uncertainties, is extended to the geotechnical model by adding values for parameters for strength, stiffness and permeability for each engineering unit. These values are found from results of tests and from understanding of basic soil behaviour.

4.6 THINGS TO DO

Ground engineering professionals should make sure that, so far as possible, they have done the standard field and laboratory tests and logged a borehole themselves. They should conduct the test, record the data and interpret these to obtain parameters.

Field work can be hazardous so it may not be possible to do all the tests. It is possible to do the standard laboratory tests such as grading, Atterberg limits, compaction, triaxial, 1D consolidation and shear. All geotechnical engineering professionals should do these tests themselves at least once, not watch someone else do them.

4.7 FURTHER READING

Most of the activities in ground investigation are specified in national codes and standards. These specify equipment for field and laboratory tests, how samples are obtained, how the test is conducted and how data are obtained. They often specify how test data are recorded and how soil parameters are derived from test data.

There are several textbooks on field and laboratory testing. These mostly give details of tests in specific national codes and standards.

Part B

Essential Physics and Mechanics

Chapter Five

A Little Basic Mechanics

Soil mechanics is the mechanics of granular materials and all the basic definitions and theories of classical mechanics apply equally to soil mechanics.

Most current theories of soil mechanics assume soil is a continuum and stress and strain vary smoothly through the ground unless a well-defined slip plane forms. Some modern particulate theories of soil mechanics consider the forces on individual grains and the interactions between them. These particulate theories are not commonly applied in ground engineering practice.

The stresses in this chapter and in Chapter 6 are total stresses and there is no distinction between total and effective stress (Section 8.1).

5.1 MASS, FORCE AND COMMON UNITS IN GROUND ENGINEERING

There are some basic principles and definitions for loading and material properties that are used for all engineering analyses.

Free fall	At rest	Units
A body with mass m in free fall accelerates at g. Mass m Acceleration g	A force F = mg will stop the body accelerating. Its weight is W = mg Weight W Force F = W = mg	Force = Newtons (N) • $1 \text{ kN} = 1000 \text{ N} = 10^3 \text{ N}$ • $1 \text{ MN} = 1000 \text{ kN} = 10^6 \text{ N}$ Mass = grams (g) • $1 \text{ kg} = 1000 \text{ g}$ • $1 \text{ Mg} = 1000 \text{ kg} = 10^6 \text{ g}$ Acceleration on Earth $g = 9.81 \text{ ms}^{-2}$ (Approximate to $g = 10 \text{ ms}^{-2}$) $100 \text{ g} \times 10 \text{ ms}^{-2} = 1 \text{ N}$

	Mass m	Weight W	Unit weight	Specific gravity G_s
1 apple	100 g = 0.1 kg	1 N		An apple floats so $G_s < 1$
1 m³ of water	1 Mg	10 kN	$\gamma_w = 10 \text{ kNm}^{-3}$	1.0
1 m³ of soil	2 Mg	20 kN	$\gamma = 20 \text{ kN}^{-3}$	For an individual soil grain $G_s \approx 2.7$

Except for some special cases such as earthquake loadings and machine vibrations, ground engineering involves stationary ground and water or water moving relatively slowly so dynamic effects and accelerations are not significant. All loads applied by the ground or by water depend on unit weights; for most soils $\gamma \approx 20 \text{ kN}^{-3}$ and for water $\gamma_w \approx 10 \text{ kNm}^{-3}$.

5.2 WHAT DO STRUCTURES HAVE TO DO?

The loads and stresses and displacements and strains in all loaded and deforming bodies have to satisfy a number of fundamental requirements.

Forces result in stresses and deformations result in strains. These are linked by:

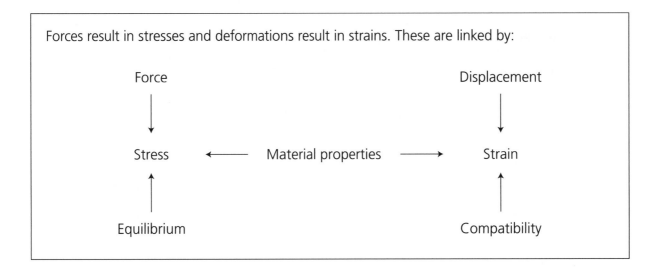

There are basic conditions that must be satisfied:

Equilibrium: the forces on anything that is not accelerating (it can be moving at constant velocity) are in equilibrium—this is simply Newton's first law. If forces are not in equilibrium the out of balance force produces an acceleration.

Compatibility: if a body moves or distorts it must move in such a way that gaps do not appear and parts do not overlap; the deformations must be compatible.

Material properties: the behaviour of the components and of the body are governed by the properties of the material. Stresses and strains are related by constitutive relationships that contain parameters describing strength and stiffness (Section 6.5).

5.2 WHAT DO STRUCTURES HAVE TO DO? (CONT)

Type of structure	Equilibrium	Compatibility
Rigid-body mechanism Moments about the pivot are in equilibrium. The beam does not bend so the angles of rotation are the same.	$F_1 L_1 = F_2 L_2$	$\dfrac{\delta_1}{L_1} = \dfrac{\delta_2}{L_2}$
Frame or structure The forces at the joint are in equilibrium because the force polygon closes. As the tension and compression members deform they remain joined.		
Continuum Forces due to stresses are in equilibrium. The square with broken lines distorts to become the trapezium.		

These basic cases apply to analyses of geotechnical structures. Rigid-body mechanisms and force polygons are used in limit equilibrium analyses (Section 15.2). Continuum mechanics is the basis of numerical modelling (Section 13).

5.3 STRESS AND STRAIN: STRENGTH AND STIFFNESS

Forces result in stresses and movements are the result of strains. The basic normal and shear stresses and strains are

	Stress	Strain
	Normal stress $\sigma = \dfrac{N}{A}$ (compression +) Shear stress $\tau = \dfrac{T}{A}$	$\delta\varepsilon_v = \dfrac{\delta L}{L_0}$ = change of size (compression +) $\delta\gamma = \dfrac{\delta H}{L_0}$ = change of shape

Basic material properties are strength which is a limiting shear stress and stiffness which is how much strain a change of stress causes. Strength and stiffness are very different properties and there is no direct relationship between them.

Basic stress–strain behaviour	Strength	Stiffness
	Strength = limiting stress Shear strength = τ_f Strong = high strength Weak = low strength	Relationship between change of stress and change of strain Tangent shear modulus $G_t \approx \dfrac{\delta\tau}{\delta\gamma}$ Secant shear modulus $G_s = \dfrac{\Delta\tau}{\Delta\gamma}$ Stiff = large modulus Soft = low modulus

5.3 STRESS AND STRAIN: STRENGTH AND STIFFNESS (CONT)

Materials may be relatively strong (Blu Tack) or relatively weak (uncooked pastry)—how easy are they to break? They may be relatively stiff (string) or relatively soft (elastic band)—how easy are they to stretch? These familiar materials have strengths and stiffnesses comparable to those of soils.

Strength and stiffness are fundamentally different and they are not related.

Changes of stress and strain:

As a structure is built and during its working life, stresses and strains in the ground change. It is important to distinguish between large and small changes.

$d\sigma$ and $d\varepsilon$ are infinitesimally small changes: $\dfrac{d\sigma}{d\varepsilon}$ = tangent stiffness

$\delta\sigma$ and $\delta\varepsilon$ are small finite changes: $\dfrac{\delta\sigma}{\delta\varepsilon} \approx$ tangent stiffness

$\Delta\sigma$ and $\Delta\varepsilon$ are large changes: $\dfrac{\Delta\sigma}{\Delta\varepsilon}$ = secant stiffness

Increments are rated by $\Delta\sigma = \sum \delta\sigma = \int d\sigma$

It is possible to determine an absolute stress but not an absolute strain. Strains are due to changes of stress from some starting point so strains can only be changes of strain related to a starting point.

5.4 ANALYSIS OF STRESS BY THE MOHR CIRCLE CONSTRUCTION

It is often necessary to calculate stresses on different planes within a loaded body.

There are stresses σ_a, σ_b, τ_{ab} and τ_{ba} on the planes A–A and B–B and stresses σ_r, σ_s, τ_{rs} and τ_{sr} on the planes R–R and S–S. The angle between the planes is θ_s. The stresses on the planes A–A and B–B are related to those on the planes R–R and S–S by equations that include the angle θ_s and satisfy equilibrium. These can be solved graphically by a Mohr circle construction. Every point on the circle represents a combination of σ and τ on a particular plane and the angle between the planes is 2θ at the centre of the circle.

There are principal planes 1–1 and 3–3 on which the shear stresses are zero and the stresses σ_1 and σ_3 are principal stresses and are the largest and smallest on any planes.

Stresses on planes	Mohr circle solution
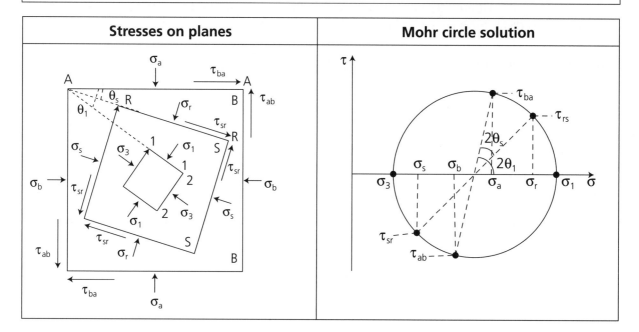	

There are some useful combinations of stress that can be represented on a Mohr circle.

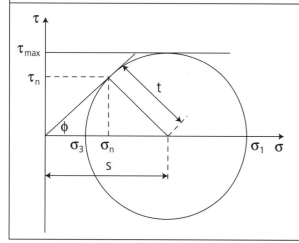

$$t = \tfrac{1}{2}(\sigma_1 - \sigma_3); \quad s = \tfrac{1}{2}(\sigma_1 + \sigma_3)$$

$$\frac{t}{s} = \sin\phi$$

$$\frac{\tau_n}{\sigma_n} = \tan\phi$$

$$\frac{\sigma_1}{\sigma_3} = \frac{1 + \sin\phi}{1 - \sin\phi} = \tan^2\left(45 + \tfrac{1}{2}\phi\right)$$

If the circle represents failing soil then τ_{max} and ϕ are alternative descriptions of its strength.

5.5 THINGS TO DO

Everything in this section appears in basic texts on structures and strength of materials. The only difference is that in geotechnical analysis compressive stresses and strains are positive while in structural analysis tensile stresses and strains are usually positive.

1. Study your books and lecture notes on structures and strength of materials and confirm that the theories and definitions in them are the same as those for geotechnical engineering here.

2. Investigate the relative strengths and stiffnesses of common materials and objects; say which are relatively strong or weak and which are relatively stiff or soft. Do simple loading tests with your hands on common materials such as a biscuit, an elastic band, warm butter, cold butter. How do these relate to the properties of common soils (Section 11.10)?

3. Do simple calculations and estimate the stresses applied by: a car tyre, your foot, a stiletto heel, the water pressure at the deep end of a swimming pool.

4. Find out how much settlement a typical house can tolerate before the owner starts to complain. Can this be converted into strains in the ground?

These exercises and tasks are important and informative. Engineers need to have a 'feeling for numbers'. They should know instinctively when a number is outside a reasonable range because if they don't that is when mistakes are made.

Chapter Six

Behaviour of Materials

The overall behaviour of materials can be examined by loading samples in different ways.

When a sample is loaded it deforms and it may 'break' if the loading reaches the strength of the material. The basic theories of cohesion, friction, elasticity and plasticity for strength and stiffness of materials are applied to soils.

The stresses in this chapter and in Chapter 5 are total stress and there is no distinction between total and effective stresses (Section 8.1).

6.1 SPECIFIC LOADING CONFIGURATIONS

There are special conditions of stress and strain that are important for soil testing and for geotechnical design and there are some special parameters for stress and strain associated with these (see also Section 9.1).

Loading condition		Stresses	Strains
Isotropic compression		σ	ε_v = volume strain
One-dimensional (1D) compression		σ_v	ε_v = vertical strain Since $\varepsilon_h = 0$ vertical and volume strains are the same
Shear		σ = normal stress τ = shear stress	ε_v = vertical strain γ = shear strain When $\gamma = 0$ this is the same as 1D compression
Axisymmetric or triaxial compression		$p = \frac{1}{3}(\sigma_a + 2\sigma_r)$ $q = (\sigma_a - \sigma_r)$	$\varepsilon_v = (\varepsilon_a + 2\varepsilon_r)$ $\varepsilon_s = \frac{2}{3}(\varepsilon_a - \varepsilon_r)$
Plane strain compression ($\varepsilon = 0$ out of the page)		$s = \frac{1}{2}(\sigma_v + \sigma_h)$ $t = \frac{1}{2}(\sigma_v - \sigma_h)$	$\varepsilon_v = (\varepsilon_v + \varepsilon_h)$ $\varepsilon_\gamma = (\varepsilon_v - \varepsilon_h)$

6.2 STIFFNESS: COMPRESSION AND DISTORTION

For each loading configuration there are stresses and strains; the strains describe both changes of size and changes of shape.

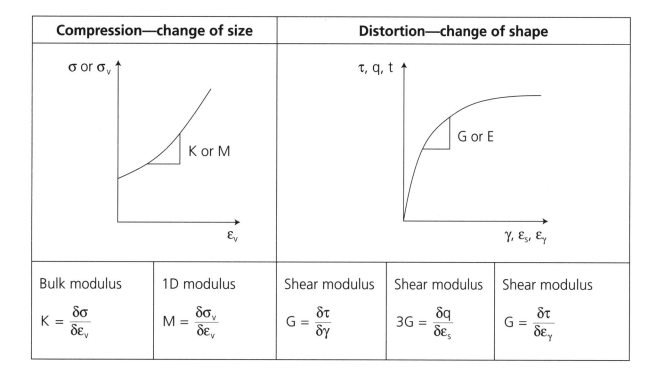

Compression—change of size		Distortion—change of shape		
Bulk modulus	1D modulus	Shear modulus	Shear modulus	Shear modulus
$K = \dfrac{\delta\sigma}{\delta\varepsilon_v}$	$M = \dfrac{\delta\sigma_v}{\delta\varepsilon_v}$	$G = \dfrac{\delta\tau}{\delta\gamma}$	$3G = \dfrac{\delta q}{\delta\varepsilon_s}$	$G = \dfrac{\delta\tau}{\delta\varepsilon_\gamma}$

Notes. The word 'compression' can have several different meanings and for clarity should have a qualification such as isotropic compression or triaxial compression.

Compression on its own means change of size. If the size increases 'negative compression' associated with decrease of mean stress, it is called swelling (Section 9.3) and if it is associated with shearing, it is called dilation (Section 11.7).

In most loading cases there will be simultaneous compression (change of size = volume strain) and distortion (change of shape = shear strain).

6.3 STRENGTH

Strength determines when a material 'breaks' under load. Common terms are tensile strength, compressive strength and shear strength. These are linked by the Mohr circle. A material fails when the Mohr circle for its stress state reaches a limiting size.

Loading condition	Stress–strain curve	Mohr circle at failure
Isotropic or hydrostatic	A fluid such as water cannot sustain a shear stress; this is why the surface of stationary water is level	
Uni-axial tension	Tensile strength	
Triaxial compression	Compressive strength	
Simple shear	Shear strength	

6.4 THEORIES TO DEFINE STRENGTH

Strength is the limiting shear stress which is described by the largest Mohr circle that the material can sustain under the current conditions of loading. Unbonded soil cannot sustain tensile stress so is only necessary to consider compressive stresses. The basic theories for soil and rock strength relate shear strength to normal stress.

Theory	Criterion	Basic formula	Mohr circles and envelope
Cohesion	Strength is independent of normal stress c = cohesive strength	$\tau = c$	
Friction	Strength increases linearly with normal stress ϕ = angle of friction	$\tau = \sigma \tan \phi$	
Mohr–Coulomb	Combined cohesion and friction	$\tau = c + \sigma \tan \phi$	
Simple Hoek–Brown	Strength increases non-linearly with normal stress	$\tau = A\sigma^b$	

6.5 STRESS–STRAIN RELATIONSHIPS

Constitutive equations

The general relationship between changes of stress and changes of strain is a constitutive equation of the form

$$\{\delta\sigma\} = [S]\{\delta\varepsilon\} \quad \text{or} \quad \{\delta\varepsilon\} = [C]\{\delta\sigma\}$$

where $\{\delta\sigma\}$ contains six stresses $\delta\sigma_x$, $\delta\sigma_y$, $\delta\sigma_z$, $\delta\tau_{xy}$ $\delta\tau_{yz}$ $\delta\tau_{zx}$ and $\{\delta\varepsilon\}$ contains six strains $\delta\varepsilon_x$, $\delta\varepsilon_y$, $\delta\varepsilon_z$, $\delta\gamma_{xy}$ $\delta\gamma_{yz}$ $\delta\gamma_{zx}$. $[S]$ is a stiffness matrix and $[C]$ is a compliance matrix.

The components of the stiffness and compliance matrices should describe the stress–strain behaviour throughout loading and unloading up to and including failure. They contain parameters that are basic soil properties and they may include the current stress and water content.

There are simple constitutive equations for the specific loading configurations described in Section 6.1 using the parameters defined there.

For axisymmetric and triaxial loading	For 1D and shear loading
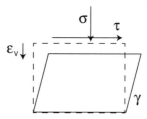	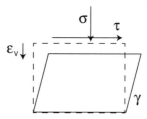
$\delta\varepsilon_s = \dfrac{1}{3G}\delta q + \dfrac{1}{J}\delta p$ $\delta\varepsilon_v = \dfrac{1}{J}\delta q + \dfrac{1}{K}\delta p$	$\delta\gamma = \dfrac{1}{G}\delta\tau + \dfrac{1}{L}\delta\sigma$ $\delta\varepsilon_v = \dfrac{1}{L}\delta\tau + \dfrac{1}{M}\delta\sigma$
G is the shear modulus, K is the bulk modulus and J is a modulus that couples shear and volumetric effects.	G is the shear modulus, M is the 1D compression modulus and L is a modulus that couples shear and volumetric effects.

Elastic and linear materials

For a material that is isotropic and elastic $J = 0$ and $L = 0$ and shear and volumetric effects are *decoupled*. This means that changes of shear stress cause only shear strains and changes of normal stress cause only volumetric strains. For isotropic elastic materials there are only two independent moduli (Section 10.8).

6.6 ELASTIC AND INELASTIC BEHAVIOUR

In a material that is elastic all strains that occur during loading are recovered if the loading is removed. The behaviour is conservative. It need not be linear.

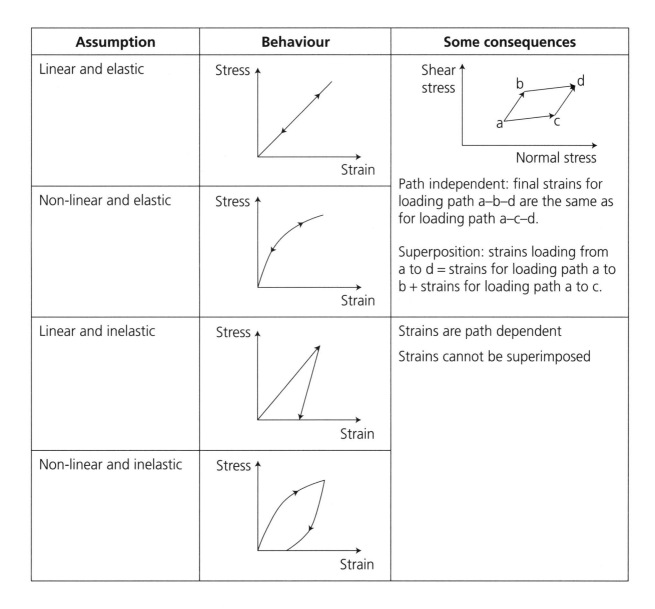

Assumption	Behaviour	Some consequences
Linear and elastic	*Stress vs Strain: straight line*	*Shear stress vs Normal stress: parallelogram a–b–d–c* Path independent: final strains for loading path a–b–d are the same as for loading path a–c–d.
Non-linear and elastic	*Stress vs Strain: concave curve*	Superposition: strains loading from a to d = strains for loading path a to b + strains for loading path a to c.
Linear and inelastic	*Stress vs Strain: triangular loop*	Strains are path dependent Strains cannot be superimposed
Non-linear and inelastic	*Stress vs Strain: hysteresis loop*	

Many routine analyses in structural and ground engineering involve adding effects and so they use the principle of superposition which is strictly valid only for elastic materials.

6.7 PERFECTLY PLASTIC BEHAVIOUR

A material that is rigid and perfectly plastic does not deform until it yields. The yield stress is the same as the failure strength and there is plastic flow. A flow rule relates combinations of plastic strain to the yield envelope. If the plastic strains are normal to the envelope the flow is associated.

Stress and strain	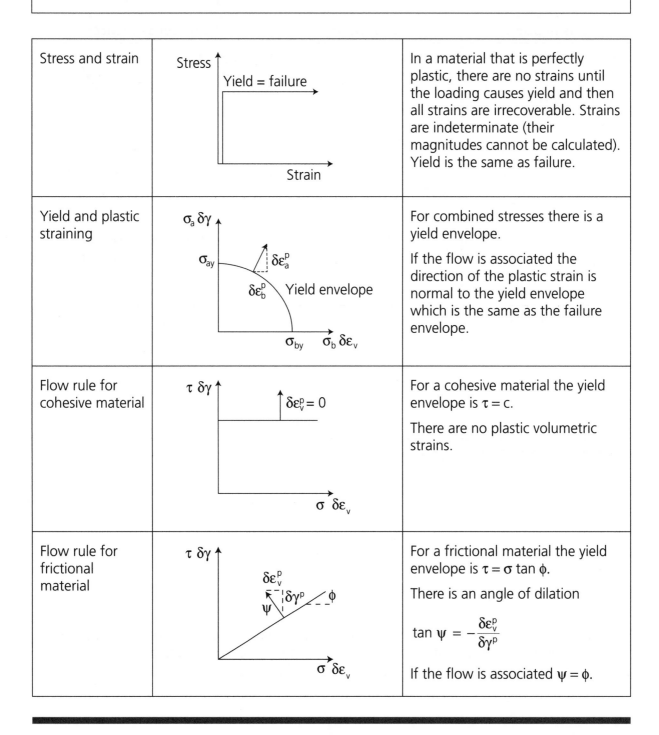	In a material that is perfectly plastic, there are no strains until the loading causes yield and then all strains are irrecoverable. Strains are indeterminate (their magnitudes cannot be calculated). Yield is the same as failure.
Yield and plastic straining		For combined stresses there is a yield envelope. If the flow is associated the direction of the plastic strain is normal to the yield envelope which is the same as the failure envelope.
Flow rule for cohesive material		For a cohesive material the yield envelope is $\tau = c$. There are no plastic volumetric strains.
Flow rule for frictional material		For a frictional material the yield envelope is $\tau = \sigma \tan \phi$. There is an angle of dilation $$\tan \psi = -\frac{\delta \varepsilon_v^p}{\delta \gamma^p}$$ If the flow is associated $\psi = \phi$.

6.8 ELASTICITY, PLASTICITY AND HARDENING

Theories of elasticity and plasticity are combined in different ways to develop models that are elastic–plastic and elasto-plastic. (Note that the difference in spelling is small but the difference in the models is considerable.)

Elastic–plastic

In a material that is elastic–plastic the strains before failure are elastic and at failure they become perfectly plastic.

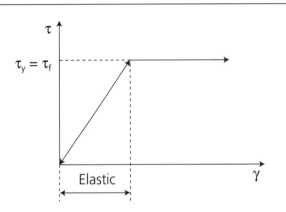

Hardening

A material is hardening when the yield stress increases with plastic strain. The hardening law relates the increase in yield stress to the plastic strain such as

$$H = \frac{\delta \tau_y}{\delta \gamma^p}$$

The maximum yield stress is the failure stress.

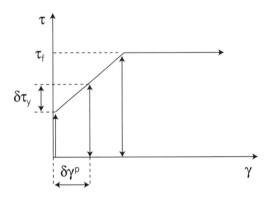

Elasto-plastic

In a material that is elasto-plastic there are simultaneous elastic and plastic strains up to failure and at failure they become perfectly plastic

$$\delta \gamma = \delta \gamma^p + \delta \gamma^e$$

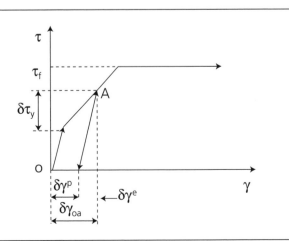

6.9 CREEP

In an elastic or an elasto-plastic material, deformations and strains arise as a consequence of changes of stress. In a perfectly plastic material that is failing, plastic strains continue at constant stress.

A material creeps when strains continue with stresses that are constant and are less than those that cause failure. Simple creep theories relate creep strains to time.

Volumetric creep

With constant isotropic or 1D stress, volumetric creep strain rate decays with time.

$$\frac{d\varepsilon_v^c}{dt} = C_\alpha \frac{1}{t}$$

and the creep strain is

$$\delta\varepsilon_v^c = C_\alpha \ln\frac{t}{t_0}$$

where t_0 is a reference time at which $\delta\varepsilon_v^c = 0$

The creep parameter C_α depends on the material and also on the current stress; creep rates are faster with larger stresses.

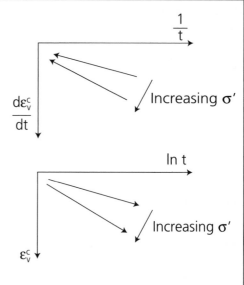

Shear creep

Shear creep with constant shear stress may decay or accelerate with time depending on the shear stress related to the strength τ_f.

If $\tau = \tau_f$ the shear strain rate is immediately large and this is plastic flow.

If τ is small compared to τ_f the shear creep rate decays with time.

There is a critical shear stress τ_c above which shear creep accelerates.

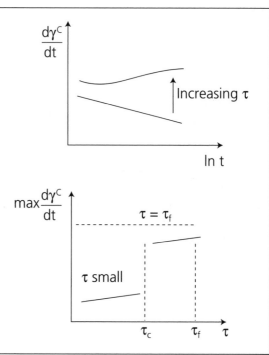

6.10 THINGS TO DO

Everything in this section appears in basic texts on structures and strength of materials.

1. Study your books and lecture notes on structures and strength of materials and confirm that the theories and definitions there are basically the same as those for geotechnical engineering here.

2. In Section 6.5 the basic constitutive equation is $\{\delta\varepsilon\} = [C]\{\delta\sigma\}$ where $\{\delta\sigma\}$ contains six stresses $\delta\sigma_x$, $\delta\sigma_y$, $\delta\sigma_z$, $\delta\tau_{xy}$ $\delta\tau_{yz}$ $\delta\tau_{zx}$ and $\{\delta\varepsilon\}$ contains six strains $\delta\varepsilon_x$, $\delta\varepsilon_y$, $\delta\varepsilon_z$, $\delta\gamma_{xy}$ $\delta\gamma_{yz}$ $\delta\gamma_{zx}$ and $[C]$ is a compliance matrix. Search your books on strength of materials and discover what the components of $[C]$ are for an isotropic and elastic material and an anisotropic and elastic material.

3. In Section 6.1, for axisymmetric compression ε_s is defined as $\frac{2}{3}(\varepsilon_a - \varepsilon_r)$. Discover why the term $\frac{2}{3}$ is required.

6.11 FURTHER READING

Most students and engineers have their favourite books on material behaviour. These are often called 'Strength of Materials' but include sections on stiffness including elasticity, plasticity, creep and viscosity and may include structural analysis as well.

Every engineer should have, and read, the two books by J. E. Gordon, *The New Science of Strong Materials: Or Why You Don't Fall Through the Floor* and *Structures: Or Why Things Don't Fall Down*. Both are published by Penguin.

Chapter Seven

Water Pressure and Seepage

Water and water pressures have a profound effect on soil behaviour and on geotechnical engineering. Pressures in the water in the pores of soil influence its strength and stiffness. Water flows through soil providing water supply but also flooding excavations. The influence of pore pressures on strength and stiffness can be the most important aspect of soil behaviour.

7.1 WATER AND WATER PRESSURE

Pore water in the soil applies pore pressure and free water applies hydrostatic pressures to walls and dams.

Pore water pressure $u = \gamma_w h_w$	Free water pressure $p_w = \gamma_w D_w$
At the water table $u = 0$ u is the same in all directions	Free water applies a total stress to the soil surface p_w is the same in all directions

Steady-state pore pressure u remains constant with time.

Excess pore pressure $\bar{u} = u_t - u_\infty$ decays with time: u_t is the current pore pressure at a time t and u_∞ is the steady-state pore pressure.

Hydrostatic	Steady-state seepage	Consolidating
u remains constant with time and the water table is level; there is no flow.	u remains constant with time but the water table is not level and hydraulic gradients cause steady-state seepage flow.	$\bar{u}$ varies with time; there is transient flow as water drains and soil compresses; after a long time $\bar{u}$ becomes zero and transient flow stops.

7.2 SATURATION

Soil is saturated when the pore spaces contain only water and it is dry when the pore spaces contain only air. Soil is unsaturated when the pore spaces contain both air or gas with pressure u_a and water with pressure u_w.

Depth in the ground	Condition	Pore pressures
	Dry	$u = 0$
	Low degree of saturation; air voids are interconnected	$u_a = 0$ and u_w depends on water content
	High degree of saturation; gas or air as individual bubbles	Pore pressures are a combination of u_w and u_a
$-h_w$	Saturated above water table	$u = \gamma_w h_w$ where h_w and u are negative
h_w	Saturated below water table	$u = \gamma_w h_w$ where h_w and u are positive

In most practical cases the depth to the water table is not great and foundations, slopes and walls are in saturated soil although this may be above the water table where pore pressures are negative.

In arid regions where rainfall is low and evaporation is high the water table is often deeper and then geotechnical construction may be in unsaturated soil.

Natural soils are very rarely perfectly dry and they usually contain small quantities of water that can apply relatively large suctions. The behaviour of dry granular materials is relevant to flow of grain from storage hoppers and sand in an hour glass.

7.3 NEGATIVE PORE PRESSURE AND SUCTION

Above the water table pore water pressures are negative (i.e. suction) due to surface tension and creation of meniscuses. The conditions are like those in an irregular capillary tube.

For a distance above the water table that depends on grain size, soil remains saturated. Above this soil becomes unsaturated and suctions depend both on grain size and degree of saturation.

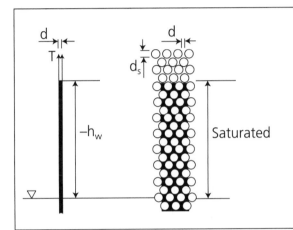

At the top of the water column and at the top of the saturated soil

$$-u_w = -\gamma_w h_w = \frac{4T}{d} = \frac{4T}{ed_s}$$

T = surface tension force, e = voids ratio and d_s = grain diameter.

The maximum suction that a soil can sustain depends on the grain size

Soil	Mean grain diameter (mm)	Maximum height of saturated soil above the water table (m)	Maximum suction (kPa)
Clay	0.001	60	600
Medium silt	0.01	6	60
Fine sand	0.1	0.6	6

In fine-grained soils (silts and clays) the soil remains saturated for considerable distances above the water table and there can be large suctions.

7.4 STEADY-STATE SEEPAGE

If pore pressures are not hydrostatic everywhere water flows through the pore spaces driven by hydraulic gradients. In steady-state seepage the flow remains constant with time.

Steady-state seepage analyses determine the quantities of flow to assess leakage through dams and the sizes of pumps to dewater excavations. Analyses also determine the variation of pore pressure throughout the region of flow to calculate effective stresses.

1D seepage through a tube of soil	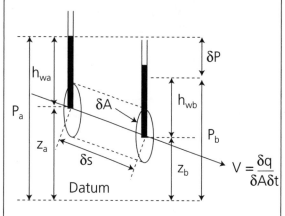	Rate of flow $\frac{\delta q}{dt}$ (m³/s) through area δA Seepage velocity $V = \frac{\delta q}{\delta t \delta A}$ (m/s) Potential $P = z + h_w = z + \frac{u}{\gamma_w}$ (m) above a datum Hydraulic gradient $i = \frac{\delta P}{\delta s}$ Darcy's law $V = ki$ (m/s) k (m/s) = coefficient of permeability
2D seepage through a simple flownet consisting of flowlines and equipotentials	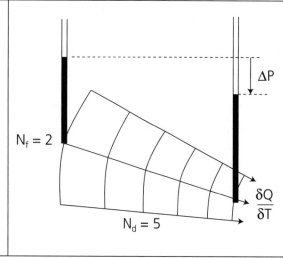	The flownet is 'square' because flowlines and equipotentials intersect at 90° and the mean length and width of each cell are the same. Rate of flow $\frac{\delta Q}{\delta t}$ (m³/s/m out of the page) through N_f flow channels with N_d potential drops $\frac{\delta Q}{\delta t} = k \frac{N_f}{N_d} \Delta P$

7.4 STEADY-STATE SEEPAGE (CONT)

Pore pressure in a flownet	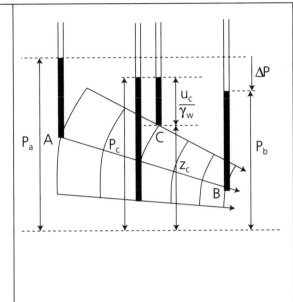	Along any flowline in a square flownet the potential drops by the same value between each equipotential.

Along any flowline in a square flownet the potential drops by the same value between each equipotential.

At an intermediate equipotential

$$P_c = P_a - \frac{N_{ac}}{N_{ab}} \Delta P$$

N_{ac} = no of drops from A to C and N_{ab} = no of drops from A to B. The pore pressure at the point C is

$$u_c = \gamma_w (P_c - z_c)$$

Flownets are important because they allow rapid and simple estimates of pore pressures which are necessary to determine effective stresses and hence soil strength and stiffness.

Leakage flow rates depend on the coefficient of permeability k which is difficult to determine. Pore pressures, however, depend only on the geometry of the flownet and can be determined much more reliably.

7.5 PERMEABILITY AND SPEED OF DRAINAGE

From Darcy's law, $V = ki$, the speed of drainage is related to the coefficient of permeability k which has the units m/s. If the hydraulic gradient is $i = 1$ then $V = k$.

The coefficient of permeability k depends on the grading of the soil and particularly on the fine fraction provided there are sufficient fines to fill the spaces between the coarse grains.

	Finest fraction	Sizes	k m/s
Coarse-grained soil	Gravel	>2 mm	$>10^{-2}$
	Sand	2–0.06 mm	10^{-2} to 10^{-5}
Fine-grained soil	Silt	0.06–0.002 mm	10^{-5} to 10^{-8}
	Clay	<0.002 mm	$<10^{-8}$

Values of k vary over orders of magnitude and it is very difficult to obtain representative values from tests on laboratory samples.

Sedimented soils are often layered with coarser and finer beds so the permeability is anisotropic and the mean permeability for vertical flow can be very different from that for horizontal flow.

In practice values for rates of seepage flow are estimated from descriptions of grading or determined from in situ pumping tests.

7.6 THINGS TO DO

There is steady-state seepage through the small sand-fill dam and through the clay flood embankment.

The sand is very permeable and the seepage velocity is fast enough to erode grains from the downstream toe.

The clay flood bank is relatively impermeable and the seepage velocity is much smaller.

Try building a small sand dam at the sea-side. What happens if the dam is over-topped?

Each tank contains saturated soil 1 m deep.

Calculate the pore water pressure at the top, bottom and at mid-height of each. (Take $\gamma = 20$ kNm^{-3} and $\gamma_w = 10$ kNm^{-3})

For the middle case water is added at the top as it drains from the bottom and it is helpful to draw a flownet. For the right-hand case the soil remains saturated.

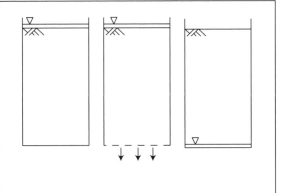

7.7 FURTHER READING

Water and water pressures are covered in books on hydraulics. These describe hydrostatics where there is no flow and flow in open channels and pipes where the flow velocity is relatively large compared with seepage velocities in soils.

Compare the Bernoulli equation in which velocity is relatively large and the potential remains constant with the theories for seepage in which the flow velocity is small and the potential drops along a flowline.

There are books that deal specifically with groundwater flow and give examples of flownets for a variety of geotechnical engineering cases.

Part C

Soil Mechanics

Chapter Eight

Pore Pressure and Effective Stress

The principle of effective stress, first formulated by Terzaghi in the 1920s, is the most important concept in soil mechanics and it is essential to understand how it works.

Effective stress and drainage of water combine to give the differences between total and effective stress analyses, drained and undrained loading and consolidation.

8.1 EFFECTIVE STRESS

This is the most important section in this book. When the total normal and shear stresses are σ and τ and the pore pressure is u the effective stresses are

$\sigma' = \sigma - u$

and

$\tau' = \tau$

The principle of effective stress says '*all effects of a change of stress such as volume change and change of strength are due to changes of effective stress*'.

Effective stresses σ' and τ' have primes and total stresses σ and τ do not. All analyses and calculations in geotechnical engineering must contain either total or effective stresses; they must not be mixed or confused.

Mohr circles of total and effective stress are the same size and are separated by a distance = u

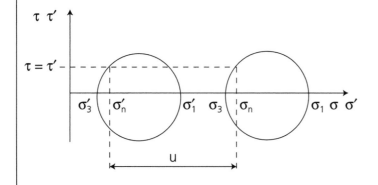

$\sigma'_1 = \sigma_1 - u$ and $\sigma'_3 = \sigma_3 - u$

Then:

$\sigma'_n = \sigma_n - u$ and $\tau' = \tau$

$s' = s - u$ and $t' = t$

$p' = p - u$ and $q' = q$

Total and effective shear stresses and deviator stresses are the same.

When pore pressures are negative the effective normal stress is larger than the total normal stress which is why unbonded soil can have an unconfined compressive strength (Section 11.5).

Warning: The principle of effective stress cannot be used in a simple way for unsaturated soil in which there are pore water pressures u_w and pore air pressures u_a (Section 19.2).

8.2 EFFECTS OF CHANGES OF STRESS

Effective stress is so called because it affects soil behaviour. If effective stress increases soil compresses (it decreases in volume and water content) and it becomes stronger; if effective stress reduces soil swells (it increases in volume and water content) and it becomes weaker.

Changes in effective normal stress are

$$\delta\sigma' = \delta\sigma - \delta u$$

so change of effective stress can be due to change of total stress $\delta\sigma$ or due to change of pore pressure δu or both simultaneously.

Increase of σ' causes settlement and soil becomes stronger	Decrease of σ' causes heave and soil becomes weaker
Foundation loading (increasing σ) with u constant	Excavation (decreasing σ) with u constant
Dewatering (decreasing u) with σ constant	Rise in groundwater (increasing u) with σ constant

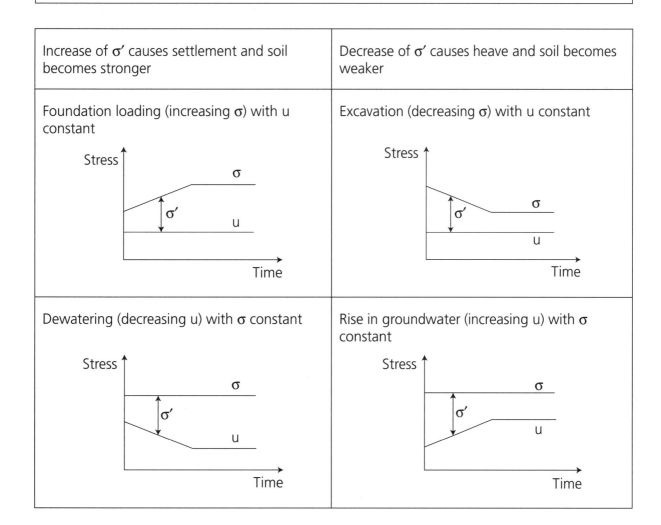

8.3 DRAINED AND UNDRAINED LOADING AND CONSOLIDATION

As saturated soil compresses or swells its water content must change and so water must seep out of or into the soil. There may or may not be enough time for drainage during construction.

The limiting conditions are fully drained loading and fully undrained loading. There is an intermediate condition of coupled loading and drainage. There is also consolidation (Sections 7.1 and 10.9) when excess pore pressures $\bar{u}$ change with constant total stress.

Routine analyses assume that the loading is either fully drained $\Delta u = 0$ or fully undrained $\Delta \varepsilon_v = 0$ or consolidating $\Delta \sigma = 0$.

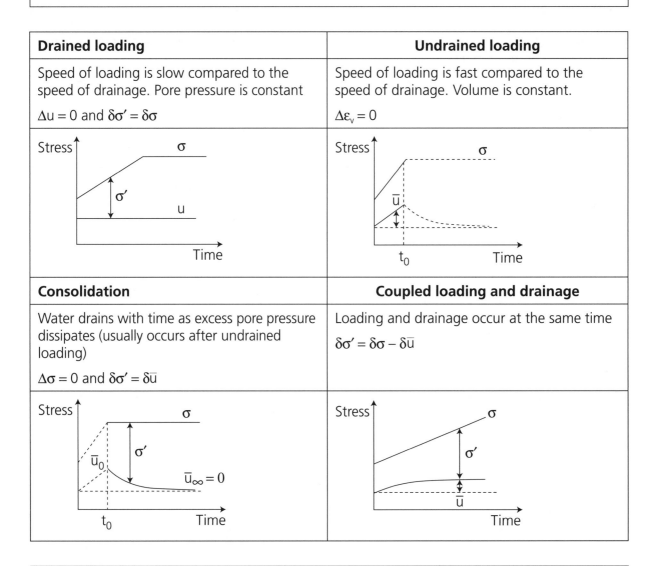

Drained loading	Undrained loading
Speed of loading is slow compared to the speed of drainage. Pore pressure is constant $\Delta u = 0$ and $\delta \sigma' = \delta \sigma$	Speed of loading is fast compared to the speed of drainage. Volume is constant. $\Delta \varepsilon_v = 0$

Consolidation	Coupled loading and drainage
Water drains with time as excess pore pressure dissipates (usually occurs after undrained loading) $\Delta \sigma = 0$ and $\delta \sigma' = \delta \bar{u}$	Loading and drainage occur at the same time $\delta \sigma' = \delta \sigma - \delta \bar{u}$

Warning: Coupled loading and drainage is not the same as coupled shearing and volume change (Section 6.5).

8.4 RATES OF LOADING

Whether the loading is drained or undrained or coupled depends on the speed of construction compared with the speed at which water can drain. These speeds vary enormously.

Construction		Drainage	
Event	Duration	Soil	Time for flow of 1 m
Earthquake	10 s	Gravel	$<10^2$ s
Small excavation	3 hours = 10^4 s	Sand	10^2 to 10^5 s
Small foundation	10 days = 10^6 s	Silt	10^5 to 10^8 s
Road embankment	3 months = 10^7 s	Clay	$>10^8$ s
Large building	3 years = 10^8 years		
Natural erosion	>30 years = 10^9 s		

For typical construction times of 10^4 to 10^8 s (3 hours to 3 years) coarse-grained soils (sands and gravels) are assumed to be drained and fine-grained soils (silts and clays) are assumed to be undrained.

But, for very fast loading during an earthquake, sand will be undrained and for very slow loading during natural erosion, clay will be drained.

Total and effective stress analyses

Routine geotechnical analyses must assume either drained or undrained loading:

Drained loading: pore pressure can be calculated, effective stresses can be found. This is an effective stress analysis.

Undrained loading: the water content does not change and pore pressures and effective stress cannot generally be calculated. This is a total stress analysis.

8.5 THINGS TO DO

Cut some sponge to fit tightly into a cylinder and saturate the sponge with water.

Apply a load to the sponge and measure how it compresses with time.

Immediately the load is applied the sponge is undrained; there is no settlement and all the load is taken by an increase in water pressure.

Then the sponge consolidates as excess water pressures dissipate.

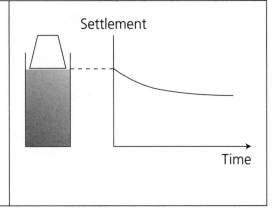

8.6 FURTHER READING

Take your favourite books on soil mechanics or ground engineering or engineering geology and study the sections on effective stress.

Do they clearly distinguish between total and effective stress and do effective stresses always have primes? In some books an effective stress is denoted $\bar{\sigma}$ but this has a different meaning to excess pore pressure $\bar{u}$.

Do all the analyses for foundations, slopes and retaining walls clearly distinguish between total and effective stress analyses and between drained and undrained loading?

Chapter Nine

Soil Behaviour Observed in Tests on Samples

The essential features of soil behaviour during loading and unloading can be seen in typical stress–strain curves from simple soil tests.

9.1 SOIL TESTING APPARATUS

The soil loading test apparatus commonly used to determine strength and stiffness parameters for routine design are triaxial cells, oedometers and direct or simple shear. Each applies stress and strain to the sample in different configurations (Section 6.1).

Apparatus	Example	Configuration
Triaxial cell: Isotropic compression $\sigma'_a = \sigma'_r$ Triaxial compression $\sigma'_a > \sigma'_r$		
Oedometer: 1D compression $\varepsilon_h = 0$		

9.1 SOIL TESTING APPARATUS (CONT)

Each apparatus applies total stresses to the sample. If u = 0 or if pore pressures are measured independently effective stresses can be determined.

Apparatus	Example	Configuration
Direct shear: Strains restricted to a thin band.		
Simple shear: Strains uniform through the sample		

9.2 LOADING OF SOIL SAMPLES

	1D compression $\varepsilon_h = 0$	Shear
Loading condition		
Stress and strain parameters and special conditions	$\varepsilon_h = 0$	When $\tau' = 0$ and $\gamma = 0$ this is the same as 1D compression
Stress path		
Simplified stress–strain behaviour		
Comment	Both loading and unloading stress–strain curves are non-linear. Strains are not recovered on unloading.	The shear stress reaches a peak and with more strain shear stress reduces but becomes constant. The soil first compresses and then dilates. At large strain the volume remains constant.

Isotropic compression $\sigma'_a = \sigma'_r$	Triaxial compression $\sigma'_a > \sigma'_r$	Notes
		All cases shown are for drained loading with $u = 0$ so $\sigma' = \sigma$
$p' = \frac{1}{3}(\sigma'_a + 2\sigma'_r)$ $\qquad q = (\sigma'_a - \sigma'_r)$ $\varepsilon_v = (\varepsilon_a + 2\varepsilon_r)$ $\qquad\quad \varepsilon_s = \frac{2}{3}(\varepsilon_a - \varepsilon_r)$		
		In triaxial compression the changes of τ' and σ' with changes of σ'_a are found from Mohr's circles
		Key features: Stress–strain curves are non-linear. Strains are not recovered on unloading. Shear stress causes volume strains; this is an example of shear-volumetric coupling.
The overall behaviour is similar to that in 1D compression.	The overall behaviour is similar to that in shear.	

9.3 CHARACTERISTICS OF SOIL BEHAVIOUR DURING COMPRESSION AND SWELLING

When soil is loaded and unloaded in isotropic or 1D compression or in the ground it compresses or swells. Volume changes are accompanied by changes of water content. Soil volume may be described by either water content or voids ratio or specific volume.

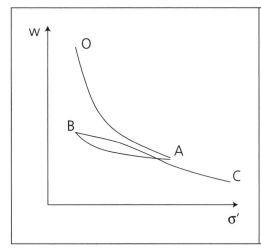

During loading and unloading the behaviour is non-linear and irrecoverable and there is hysteresis (the unloading–reloading curves have a loop).

O to A = loading or deposition

A to B = unloading or erosion in the ground

B to C = reloading or redeposition in the ground

Soil compression is mostly due to re-arrangement of grains; in addition clay grains may distort and coarse grains may fracture. When coarse soil is unloaded there is very little swelling; grains do not 'un-rearrange' or 'un-fracture'. When clay grains are unloaded they may 'un-distort' and the soil swells.

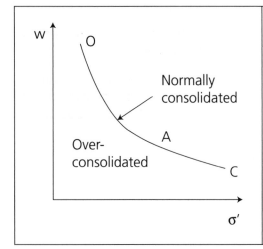

OAC is the normal consolidation line. It is a boundary between possible and impossible states for unbonded soil.

For states on OAC the soil is normally consolidated

For states to the left of OAC the soil is overconsolidated

The term *normal consolidation line* is a misnomer; it actually describes compression, the relationship between stress and volume.

9.4 CHARACTERISTICS OF SOIL BEHAVIOUR DURING SHEARING

The fundamental features of soil behaviour can be seen during drained and undrained shear tests on soils that have been compressed and swelled to different initial stresses and water contents.

1D Compression and Swelling

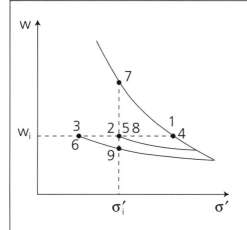

Soil behaviour depends on its initial state which is the combination of effective stress and water content (Section 12).

Soil can reach different initial states by loading and unloading in a laboratory test or by deposition and erosion in the ground.

Samples 1, 4 and 7 are normally consolidated and the remainder are overconsolidated because they have been unloaded from a NC state.

Drained and undrained shearing (Section 8.3)

Some samples are sheared drained and others are sheared undrained and they have different behaviour.

Undrained: there is no drainage. The volume and water content remain constant but the pore pressure may change.

Drained: the pore pressure remains constant. The soil may compress or dilate: this is coupled shear and volumetric behaviour.

Each of the samples 1 to 9 is sheared by increasing total shear stress τ with constant total normal stress σ. The differences between the samples and tests are:

Drainage: samples 1, 2 and 3 are sheared undrained; the remainder are sheared drained.

Initial states: samples 1 to 6 have the same initial water content w_i; samples 7, 8 and 9 have the same initial effective stress σ'_i.

In the following the data are shown with several different axes. To fully understand soil behaviour it is necessary to consider all these diagrams.

9.4 CHARACTERISTICS OF SOIL BEHAVIOUR DURING SHEARING (CONT)

Undrained shearing: all samples have the same initial water content w_i but different initial stresses σ_i'.

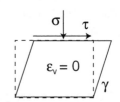

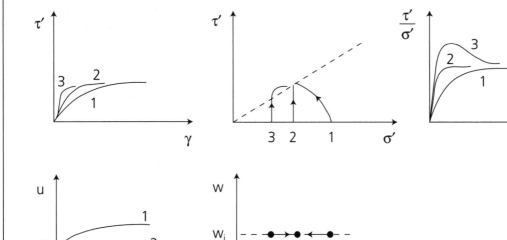

All samples are undrained so their water contents remain constant.

Pore pressures change as all samples move towards a common state.

All samples move towards the same strength τ' and the same stress ratio τ'/σ'.

9.4 CHARACTERISTICS OF SOIL BEHAVIOUR DURING SHEARING (CONT)

Drained shearing: all samples have the same initial water content w_i but different initial stresses σ'_i.

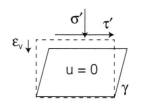

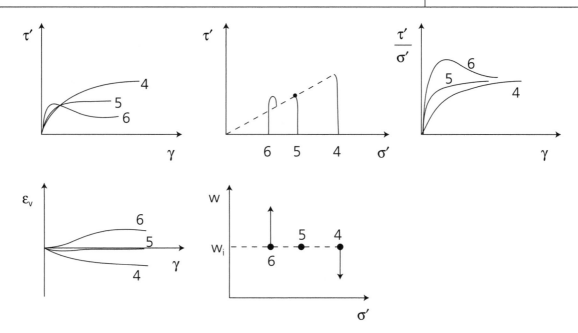

All samples are drained so their effective normal stresses remain constant.

There are volumetric strains; sample 4 compresses and sample 6 dilates (expands).

All samples move towards the same stress ratio.

9.4 CHARACTERISTICS OF SOIL BEHAVIOUR DURING SHEARING (CONT)

Drained shearing: all samples have the same initial effective stress σ'_i but different initial water contents w_i.

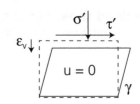

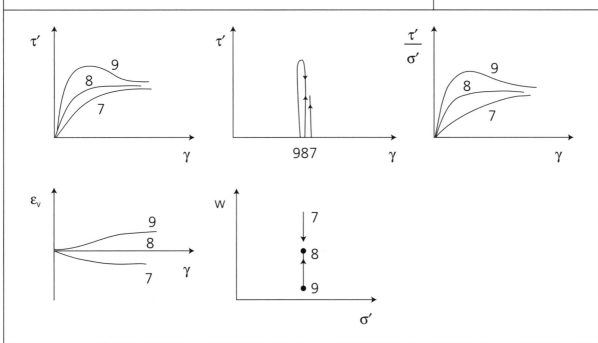

All samples are drained so their effective normal stresses remain constant.

There are volumetric strains; sample 7 compresses and sample 9 dilates (expands).

All samples move towards the same strength τ' and the same stress ratio τ'/σ'.

9.5 UNIFIED SOIL BEHAVIOUR

There is a consistent pattern in the behaviours shown in Section 9.4 that can be collected together. In the diagrams below N means normally consolidated and O means overconsolidated (Section 10.2); subscript d means drained and subscript u means undrained.

Stress–strain behaviour	State paths

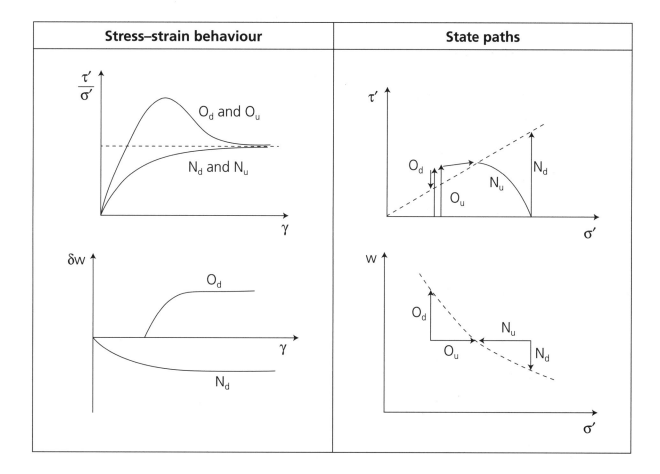

Comments

Curves of stress ratio τ'/σ' with shear strain γ are the same for drained and undrained shearing and all samples approach the same stress ratio at large shear strains.

Normally consolidated samples compress when sheared drained and the effective normal stress decreases when sheared undrained. Overconsolidated samples dilate when sheared drained and the effective stress increases when sheared undrained.

The states approach a common relationship between shear stress, normal stress and water content indicated by broken lines.

9.6 THINGS TO DO

Apparatus and tests

In a university or commercial laboratory examine the apparatus in Section 9.1. Take them to pieces yourself.

Discover exactly how the sample is installed in the apparatus, how the stresses are applied, how the measurements of stress, strain, pore pressure and volume are made.

Patterns of behaviour

Examine the behaviour of soils 1–9 in Section 9.4 and relate these to the common behaviour in Section 9.5. Consider why all the samples seem to be approaching a common state irrespective of their starting points.

Shearing and volume change

Do a simple thought experiment. Consider a dense packing and a loose packing of spheres subjected to shearing and explain why the volume of a sample changes during drained shearing; why does the dense packing dilate and the loose packing compress? Why do pore pressures change during undrained shearing?

9.7 FURTHER READING

The tests that are used in practice to investigate soil behaviour and to measure design parameters are determined by national codes and specifications.

There are books that give detailed commentaries on soil strength and stiffness testing relevant to practice in several regions of the world.

When you examine soil test data be careful to assess the probable errors in the results (Section 14.3); it is actually quite difficult to do good soil tests and many sets of data contain errors and flaws. If the sample has developed a well-defined slip-plane measurements of stress and strain in the distorted sample become unreliable.

Chapter Ten

Soil Deformation

As soil is loaded or unloaded it strains and deforms by distortion (change of shape) and, if drained, by compression or swelling (change of size). Change of stress and change of strain are related by a constitutive equation (Section 6.5) containing stiffness parameters.

For materials such as steel and concrete which are assumed to be linear and elastic there are only two stiffness parameters. These can be Young's modulus E and Poisson's ratio ν or shear modulus G and bulk modulus K. Soil behaviour is not simply linear and elastic and, for a particular soil, stiffness parameters depend on the loading configuration, the current state and the strain.

For each of the common loading configurations (Section 9.2) there are characteristic stress–strain curves and the gradients of these are stiffness parameters.

10.1 STIFFNESS OF SOIL FOR DRAINED AND UNDRAINED LOADING

	1D compression	Shear
Stress–strain curve		
Drained stiffness parameters	$M' = \dfrac{\delta\sigma'_v}{\delta\varepsilon_v}$ or $m_v = \dfrac{1}{M'}$	$G' = \dfrac{\delta\tau'}{\delta\gamma}$
Undrained stiffness parameters	$M_u = \infty$ or $m_v = 0$	$G_u = \dfrac{\delta\tau}{\delta\gamma}$
Non-linear stiffness	M' varies with stress and strain	G' decreases with strain
Combined compression and shearing		$\delta\gamma = \dfrac{1}{G'}\delta\tau' + \dfrac{1}{L'_1}\delta\sigma'$ $\delta\varepsilon_v = \dfrac{1}{L'_2}\delta\tau' + \dfrac{1}{M'}\delta\sigma'$ For isotropic and elastic soil $1/L' = 0$ and $G_u = G'$

Isotropic compression $\sigma'_a = \sigma'_r$	Triaxial compression $\sigma'_a > \sigma'_r$ and $\Delta\sigma'_r = 0$	
Isotropic compression and triaxial compression tests are done in the same apparatus		The same triaxial test results can be plotted to give either Young's modulus E or the shear modulus G
$K' = \dfrac{\delta p'}{\delta \varepsilon_v}$	$3G' = \dfrac{\delta q'}{\delta \varepsilon_s}$	$E' = \dfrac{\delta q'}{\delta \varepsilon_s}$ and $\nu' = -\dfrac{\delta \varepsilon_r}{\delta \varepsilon_a}$
$K_u = \infty$	$3G_u = \dfrac{\delta q}{\delta \varepsilon_s}$	$E_u = \dfrac{\delta q}{\delta \varepsilon_a}$ and $\nu_u = 0.5$
K' varies with stress and strain	 G' decreases with strain	 E' decreases with strain
	$\delta\varepsilon_s = \dfrac{1}{3G'}\delta q' + \dfrac{1}{J'}\delta p'$ $\delta\varepsilon_v = \dfrac{1}{J'}\delta q' + \dfrac{1}{K'}\delta p'$ For isotropic and elastic soil $1/J' = 0$	$\delta\varepsilon_a = \dfrac{1}{E'}\left[\delta\sigma'_a - 2\nu'\delta\sigma'_r\right]$ $\delta\varepsilon_r = \dfrac{1}{E'}\left[\delta\sigma'_r(1 - \nu') - \nu'\delta\sigma'_a\right]$

10.2 COMPRESSION AND SWELLING, YIELDING AND OVERCONSOLIDATION

The basic behaviour of soil during compression and swelling is described in Section 9.3 where water content was related to the effective stress σ'.

The state of a soil is the combination of stress and water content (see Chapter 12).

Along OAC the soil is normally consolidated. For any state such as at B to the left of OAC the soil is overconsolidated.

The state of unbonded soil cannot exist outside OAC.

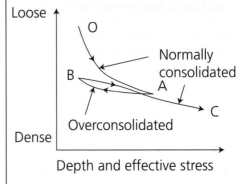

When stress is plotted to a log scale the normal consolidation line (NCL) becomes approximately linear.

When soil is normally consolidated its state is on the NCL and its behaviour is plastic.

When soil is overconsolidated its behaviour is usually simplified to elastic and non-linear (but linear with stress on a log scale).

On reloading from B the soil yields at A on the NCL.

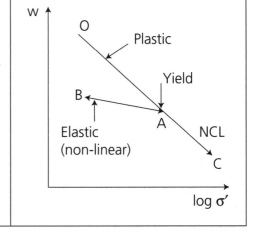

When the state is at B the yield stress is σ'_y and the maximum past stress is σ'_m.

The state at B is defined by:

Overconsolidation ratio $R_0 = \dfrac{\sigma'_m}{\sigma'} \geq 1$

or yield stress ratio $R_y = \dfrac{\sigma'_y}{\sigma'} \geq 1$

If the soil does not creep $R_0 = R_y$ but if the soil creeps and compresses at constant σ' the values of R_0 and R_y differ (Section 12.5).

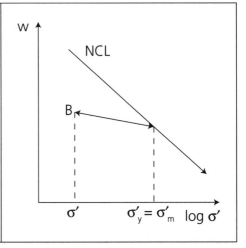

10.3 SIMPLE MODEL FOR 1D COMPRESSION

The conventional methods of recording 1D compression show voids ratio and $\log \sigma'_v$ or specific volume and $\ln \sigma'_v$.

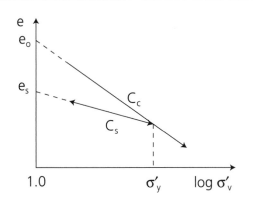

The NCL is taken as linear with slope C_c. The unloading and reloading loop is approximated to a swelling line with slope C_s

On the NCL $e = e_o - C_c \log \sigma'_v$

On a swelling line $e = e_s - C_s \log \sigma'_v$

$$\delta \varepsilon_v = -\frac{\delta e}{1 + e}$$

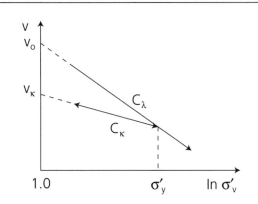

On the NCL $v = v_o - C_\lambda \ln \sigma'_v$

On a swelling line $v = v_\kappa - C_\kappa \ln \sigma'_v$

$$\delta \varepsilon_v = -\frac{\delta v}{v}$$

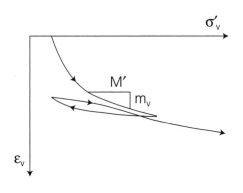

At any point on the stress–strain curves the 1D modulus is

$$M' = \frac{1}{m_v} = \frac{d\sigma'_v}{d\varepsilon_v}$$

$$\delta \varepsilon_v = -\frac{\delta v}{v} = C_\lambda \frac{\delta \sigma'_v}{v \sigma'_v} \quad \text{or} \quad C_\kappa \frac{\delta \sigma'_v}{v \sigma'_v}$$

$$M' = \frac{v \sigma'_v}{C_\alpha} \text{ or } \frac{v \sigma'_v}{C_\kappa}$$

The 1D modulus M' depends on the current stress and specific volume and on the history of loading and unloading.

10.4 GENERAL MODEL FOR ISOTROPIC AND 1D COMPRESSION

The general model for compression and swelling of soil combines isotropic and 1D states and shows specific volume v and mean stress p'.

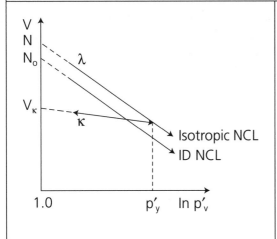

Isotropic normally consolidated states

$$v = N - \lambda \ln p'$$

1D normally consolidated states

$$v = N_o - \lambda \ln p'$$

Overconsolidated states

$$v = v_\kappa - \kappa \ln p'$$

$$\delta \varepsilon_v = -\frac{\delta v}{v} = \lambda \frac{\delta p'}{vp'} \quad \text{or} \quad \kappa \frac{\delta p'}{vp'}$$

$$K' = \frac{vp'}{\lambda} \text{ or } \frac{vp'}{\kappa}$$

Bulk modulus K' depends on the current stress and specific volume and on normally consolidated λ or swelling κ.

This basic model also includes soil failure at its critical state (Section 11). The parameters λ, N and N_0 describe the isotropic and 1D normal consolidation lines and the parameter κ gives the slope of any swelling line. These are material parameters; they depend only on the grains and are independent of state.

10.5 HORIZONTAL STRESS IN THE GROUND

When soil in the ground is loaded and unloaded by soil above being deposited or eroded there are no horizontal strains and the state is one-dimensional or 1D.

The horizontal and vertical effective stresses are related by a coefficient of earth pressure at rest $K_0 = \sigma'_h/\sigma'_v$. Empirical relationships link K_0 to soil strength given by the angle of friction ϕ' (Chapter 11) and overconsolidation described by the yield stress ratio.

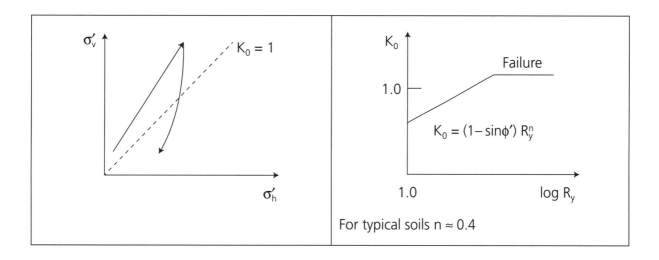

For typical soils $n \approx 0.4$

10.6 NON-LINEAR STIFFNESS OF SOIL

Shear modulus and Young's modulus decrease with shear strain. The bulk modulus decays with volumetric strain for small strain but at larger strain it increases as the soil compresses and the water content decreases.

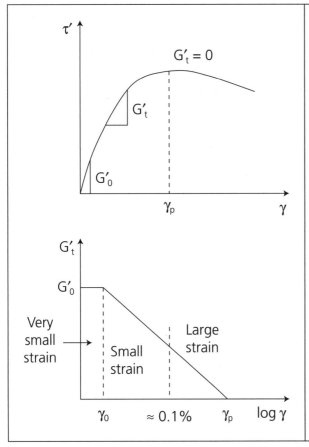

A stress–strain curve should be converted into a stiffness–strain decay curve.

At very small strain the shear modulus is G'_0 and soil is linear and elastic.

There is a strain γ_0 beyond which the stiffness decays with strain; γ_0 is very small and is normally around 0.001%.

At the peak $G'_t = 0$. The shear strain γ_p at the peak strength depends on the soil.

Conventionally the distinction between small and large strain is around 0.1%.

The diagram shows a simple linear decay of stiffness with log strain but often other empirical curves are used.

It is very difficult to measure strains and stiffnesses in routine shear and triaxial tests at strains smaller than about 0.1%. Small strains can only be measured reliably using special gauges.

The stiffness at any strain can be estimated from simple measurements and observations. The stiffness at very small strain G'_0 is related to the velocity of a shear wave (Section 10.7).

When the shear stress is the peak strength $G'_t = 0$. The strain at the peak γ_p can be measured in routine tests; it is often between 2% and 5% but can be more or less.

A linear decay of stiffness with the logarithm of strain is a reasonable approximation to the behaviour of many soils.

10.7 VERY SMALL STRAIN STIFFNESS G'_0

The very small strain shear modulus G'_0 is related to the velocity V_s of a shear wave and the unit weight of the soil γ by

$$G'_0 = \frac{\gamma V_s^2}{g}$$

V_s is relatively easy to measure in laboratory tests or in the ground.

Shear waves generated at ground level or at depth.	Vibrating bender elements in the top and bottom of a triaxial sample.

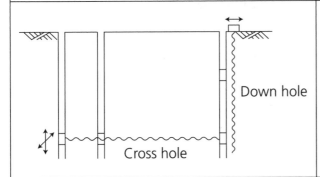

Down hole

Cross hole

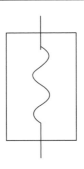

For a particular soil G'_0 depends on the current mean stress p' and on the current state. Typical values range from 20 MPa for soft soil to over 100 MPa for stiff soils.

G'_0 is an important soil stiffness parameter. It can be measured in situ or in laboratory tests and it 'anchors' the stiffness decay curve at the very small strain end.

10.8 CONSOLIDATION

When a total stress $\Delta\sigma_v$ is applied quickly fine-grained soil is undrained. Because $\Delta\varepsilon_v = 0$ and $\Delta\sigma'_v = 0$ pore pressure will increase by $\Delta u = \Delta\sigma_v$ and this is the initial excess pore pressure $\bar{u}_0$.

As time passes the excess pore pressures dissipate with constant total stress as water flows towards a drain where $\bar{u} = 0$ and the soil compresses. This is consolidation.

After a long time excess pore pressures become $\bar{u} = 0$. From the start to the finish of consolidation $\Delta\sigma'_v = \Delta\bar{u} = \Delta\sigma_v$.

Excess pore pressure in 1D consolidation

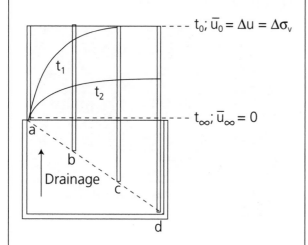

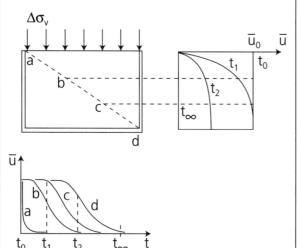

The base and sides of the container are impermeable so water can only drain upwards so drainage and deformations are 1D.

At time t_0 there is an excess pore pressure $\bar{u}_0$ everywhere as shown by the rise of water in the standpipes.

The excess pore pressure $\bar{u}$ at any depth decays with time. After a very long time $\bar{u}_\infty = 0$ everywhere.

At any point in the ground $\delta\sigma' = \delta\bar{u}$ during an interval of time δt.

Excess pore pressures decay at different rates at different depths depending on the distances from the drainage surface. The pore pressure at a point near the surface decays very quickly; lower down pore pressures do not start to decay immediately.

10.8 CONSOLIDATION (CONT)

Settlement due to consolidation

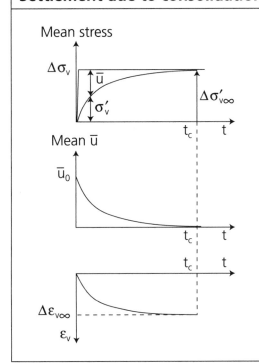

During consolidation the average excess pore pressure decreases and the average effective stress increases: since σ remains constant

mean $\delta\sigma' = $ mean $\delta\bar{u}$.

After a long time $\bar{u}_\infty = 0$ when $t = t_c$ and

$$\Delta\varepsilon_\infty = m_v\Delta\sigma' = m_v\Delta\sigma$$

Rate of Consolidation

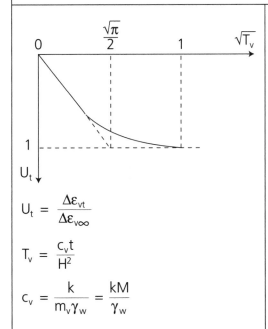

$$U_t = \frac{\Delta\varepsilon_{vt}}{\Delta\varepsilon_{v\infty}}$$

$$T_v = \frac{c_v t}{H^2}$$

$$c_v = \frac{k}{m_v\gamma_w} = \frac{kM}{\gamma_w}$$

The rate of consolidation is governed by the speed at which water can drain. This is determined by the permeability and the compressibility of the soil which together give the coefficient of consolidation c_v. The units of c_v are m²/year.

The rate of consolidation is also governed by the drainage path length H which is how far the water has to travel (Section 15.6).

Standard solutions for consolidation relate the degree of consolidation U_t to the time factor T_v.

When $U_t = 1$ and $T_v \approx 1$ the time for consolidation to be complete is

$$t_c \approx \frac{H^2}{c_v}$$

10.9 RELATIONSHIPS BETWEEN STIFFNESS MODULI FOR ISOTROPIC ELASTIC SOIL

> For an isotropic and elastic soil there are only two basic stiffness parameters and these can be G′ and K′ or E′ and v′. There are relationships between these and the corresponding undrained elastic parameters.

	Drained with $v' = 0.25$	**Undrained $v_u = 0.5$**
$G' = \dfrac{E'}{2(1+v')}$	$G' = 0.4E'$	$G_u = G'$
$K' = \dfrac{E'}{3(1-2v')}$	$K' = 0.67E'$	$K_u = \infty$
$M' = \dfrac{E'(1-v')}{(1+v')(1-2v')}$	$M' = 1.2E'$	$M_u = \infty$
$E_u = \dfrac{3E'}{2(1+v')}$	$E_u = 1.2E'$	E_u
$G' = K'\dfrac{3(1-2v')}{2(1+v')}$	$G' = 0.6\,K'$	$G_u = G'$

10.10 CREEP

Volumetric creep (Section 6.9) is compression at constant effective stress and so volumetric creep is a departure from the principle of effective stress (Section 8.1). Creep occurs particularly in soft soils with high organic content and can make a significant contribution to long-term settlement of embankments.

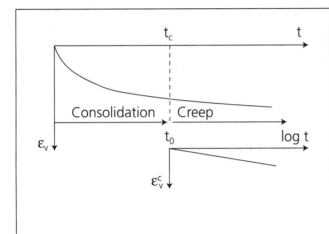

The conventional assumption is that creep starts at a time t_c when consolidation is complete and excess pore pressures have fully dissipated.

Volumetric creep strains are

$$\frac{d\varepsilon_v^c}{dt} = C_\alpha \frac{1}{t}$$

$$\Delta\varepsilon_v^c = C_\alpha \ln\left(\frac{t}{t_o}\right)$$

C_α is the coefficient of creep. It depends on the soil and on the stresses that are causing the creep.

Notes:

1. Volume changes in soil that occur before and after excess pore pressures have fully dissipated are often called primary consolidation and secondary consolidation, respectively. These are misnomers; settlements with dissipation of excess pore pressure are *consolidation* and settlements with constant pore pressure and constant effective stress are *creep*.

2. Volume changes due to consolidation may be compression if excess pore pressures generated by loading are positive or swelling if excess pore pressures generated by unloading are negative. Volume changes due to creep at constant total and effective stress can only be compression.

3. Creep changes the state of soil by reducing its volume and water content (Section 12.5) and so yield stress ratio R_y is a better description of soil state than overconsolidation ratio R_0.

10.11 THINGS TO DO

1. Do 1D compression tests on a compressible material such as corn flakes or sponge rubber. Put the sample into a straight-sided container (not glass) with a wood piston at the top. Add and remove load and measure the change of sample height. Plot height or volume strain against stress with natural and log scales.

2. Do unconfined compression tests (triaxial tests with $\sigma_r = 0$) on soft materials such as warm butter and playdough. Apply an axial stress and measure the axial strain. Determine values for Young's modulus and shear modulus G.

For each test devise simple equipment and simple methods to make cylindrical samples and to measure load and deformation. For deformations use a ruler. For loading add kitchen weights or place the sample on kitchen scales, pushdown on the top of the sample and observe the load recorded on the scales.

10.12 FURTHER READING

Most textbooks cover soil stiffness in 1D compression, shear and triaxial tests but in a variety of different ways and using several different terminology, symbols and parameters.

Most texts cover soil compression in sections titled consolidation which can be misleading. Compression is change of volume related to change of effective stress which is drained loading; consolidation is change of volume due to change of pore pressure which is a dissipation process (Section 10.9).

Chapter Eleven

Soil Strength

Strength is the limiting shear stress that soil can safely resist with the current normal stress and water content. This depends on whether the soil is drained or undrained and also on the strain and deformation.

11.1 SOIL HAS SEVERAL STRENGTHS

Typical shear stress—shear strain curves have a peak strength and a critical state strength. If the shearing is drained there are volumetric strains. After large deformations some soils have a smaller residual strength.

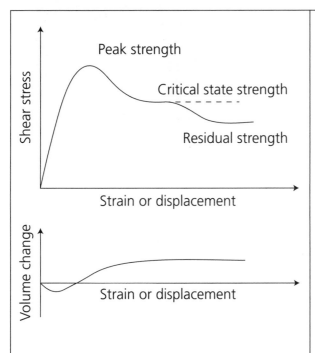

Peak strength

This may be either the maximum shear stress τ'_p or the maximum stress ratio $(\tau'/\sigma')_p$ which may occur at the same strain or different strains. Strains at the peak are relatively small (of the order of 1%). At the peak strength the volume is increasing (the soil is dilating).

Critical state strength

When soil is at its critical state it continues to distort at constant effective stress and constant volume.

Residual strength

When clay soils form slip planes they may reach a residual strength lower than the critical state strength.

Residual strength and critical state strength are different phenomena. At the critical state grains are rotating and rearranging and the flow is turbulent; at the residual flat clay grains slip one over another and the flow is laminar. Many books do not distinguish between these two very different processes.

In practice, the residual strength only matters in two cases. First, when there is a pre-existing slip plane in the ground usually the result of previous landsliding. Second, when construction results in very large deformations usually along the sides of driven piles (Section 17.7).

11.2 CRITICAL STATE STRENGTH

Soils reach critical states when they continue to deform at constant effective stress and constant volume without a well-defined slip plane.

The sand in the continuously rotating cylinder is at its critical state; the slope angle and height remain constant.

All that happens is the grains are continuously rearranging.

The angle of the slope is the same as the critical state friction angle ϕ'_c (Section 16.3).

Logically, granular materials **must** reach critical states at which they continue to deform at constant stress and constant volume. If they did not the slope in the rotating cylinder would gradually become steeper or flatter and the volume of soil would continue to increase or decrease—both possibilities are nonsense.

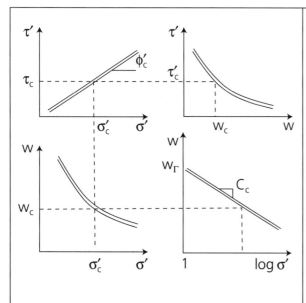

When soil is at its critical state there is a unique relationship between shear stress, normal stress and water content that is independent of the initial starting state.

It is conventional to show the critical state line on a diagram by a double line.

$$\tau'_c = \sigma'_c \tan\phi'_c$$

$$w_c = w_\Gamma - C_c \log\sigma'_c$$

$$\tau'_c = \tan\phi'_c 10^{\frac{w_\Gamma - w_c}{C_c}}$$

The critical state parameters ϕ'_c, C_c and w_Γ are material parameters; they depend only on the grains.

11.3 RESIDUAL STRENGTH

In soil at its critical state the grains are rotating and rearranging themselves. In soil at its residual state flat platy clay grains have become aligned in thin zones and as the soil deforms clay platelets slip one on another.

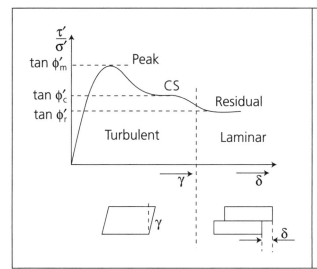

Critical state deformations are turbulent; residual deformations are laminar. The residual friction angle ϕ_r' can be very much smaller than the critical friction angle ϕ_c'.

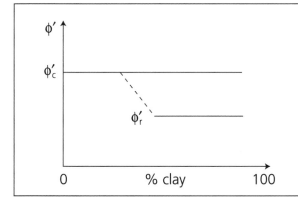

Soils with grains that are rotund (not flat and platy) do not have a residual strength that is smaller than the CS strength. Well-graded soils need more than about 40% clay to have residual strength.

The surface of a residual slip zone appears shiny. You are looking at the surfaces of clay platelets that have been aligned parallel with the slip surface.

(Photograph from Skempton, A. W. 1964. Long-term stability of clay slopes. *Geotechnique* 14(2), 77–101.)

11.4 UNDRAINED STRENGTH AND EFFECTIVE STRESS STRENGTH

Critical state strength depends on water content and effective normal stress. If soil is undrained the water content does not change so the critical state strength depends on the initial water content and is the same for all total normal stresses. At the same time the strength is related to the effective normal stress through an angle of friction.

Loading condition	Strength model	Strength equation	Mohr circles
Undrained loading of saturated soil	Cohesive	$\tau = s_u$ Many books use c_u but it is better to use s_u to mean undrained strength.	
Any loading for which effective normal stress is known – this is mostly when the soil is drained	Frictional	$\tau' = \sigma' \tan \phi'_c$	

These two strength models are complementary. If the soil is undrained use $\tau = s_u$. If pore pressures are known, usually when soil is drained, use $\tau' = \sigma' \tan \phi'_c$. Never mix the two.

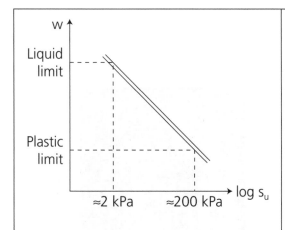

At the liquid limit the soil is nearly liquid and at the plastic limit it is stronger and brittle. These states correspond to fixed strengths of approximately 2 and 200 kPa.

Undrained strength is related to liquidity index I_L (Section 3.8) which defines where the water content is between the liquid limit and the plastic limit by

$$s_u = \frac{200}{10^{2I_L}}$$

This is a useful method for determining an approximate value for undrained strength (Section 14.6).

11.5 UNCONFINED COMPRESSIVE STRENGTH

Both coarse-grained sand and fine-grained clay have an unconfined compressive strength; neither are bonded and both have effective stress frictional strength.

The fine-grained clay is much stronger than the coarse-grained sand because it can generate a much larger pore water suction.

The Mohr's circles are geometrically the same for both soils.

The total and effective stress Mohr's circles are always the same size.

The total stress circle is defined by the unconfined compressive strength (UCS) σ_a and $\sigma_r = 0$

The effective stress circle is limited by the failure envelope $\tau' = \sigma' \tan \phi'$

The circles are separated by the pore pressure u which is negative

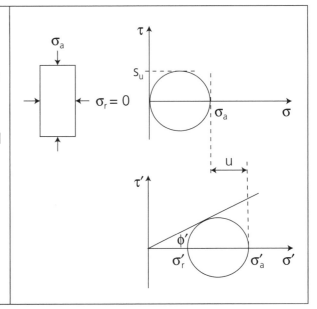

In total stress there is an undrained shear strength $s_u = \frac{1}{2}\sigma_a = \frac{1}{2}\text{ucs}$

Unconfined compressive strength of unbonded soil arises from negative pore pressure that generates a positive effective normal stress that gives soil a shear strength due to friction.

The negative pore pressure that a soil can generate depends on its grain size: fine-grained soil can generate much larger suctions than coarse-grained soil (Section 7.3). This is why the clay is stronger than the sand.

11.6 PEAK STRENGTH OF SOIL

Soils that are overconsolidated have a strength at small strain that is larger than the critical state strength. These soils dilate when they shear at constant effective normal stress; this is an example of shear and volumetric coupling.

Drained shearing of soil starting from the same water content and different stresses (Section 9.4).

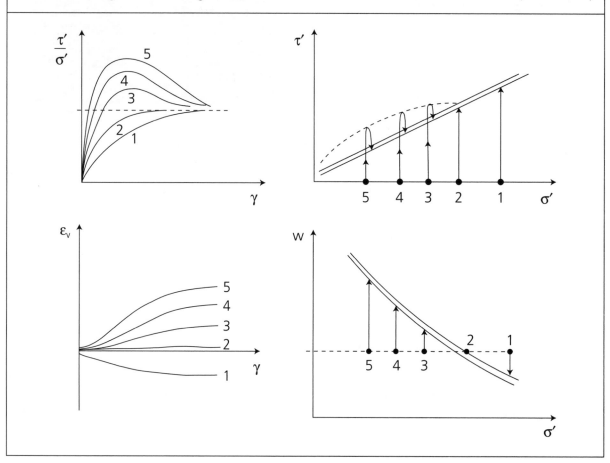

11.6 PEAK STRENGTH OF SOIL (CONT)

Drained shearing of soil starting from the same stress and different water content (Section 9.4).

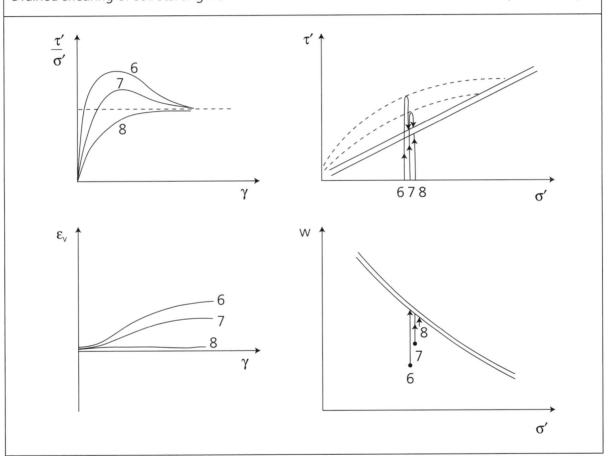

The peak strengths are limited by a curved envelope. The size of the envelope increases as the initial water content decreases.

11.7 STRESS DILATANCY

When soil is at its peak strength it is dilating; the shear stress has to overcome friction and raise the normal stress. This is analogous to a frictional block being pushed up a plane. If $i = \psi$ the direction of movement of the forces T and N is the same as that of the stresses τ' and σ'.

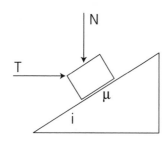	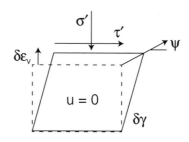
For equilibrium with the block moving up the plane $$\frac{T}{N} = \tan(\mu + i)$$	The mechanics of the dilating soil are the same as those of the sliding block $$\frac{\tau'}{\sigma'} = \tan(\phi'_c + \psi) = \tan\phi'_m$$

The basic stress–dilatancy relationship applies throughout the whole loading not just at the peak. When $\psi = 0$ then $\phi' = \phi'_c$ and at the peak when $\psi = \psi_m$ then $\phi' = \phi'_m$.

11.7 STRESS DILATANCY (CONT)

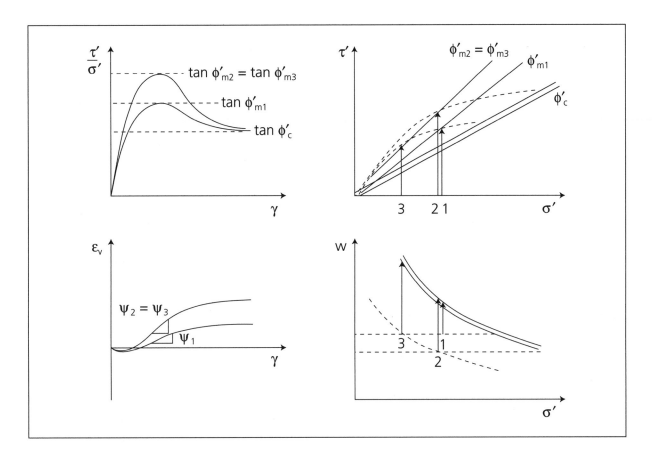

Samples 1 to 3 have different initial states and they all have the same critical state stress ratio $\tan \phi'_c$.

Samples 1 and 2 have the same effective normal stress but different initial water contents. They dilate by different Δw; they have different peak stress ratios $\tan \phi'_{m1} \neq \tan \phi'_{m2}$.

Samples 1 and 3 have the same water content but different effective normal streses. Their peak strengths are on the same curved envelope (Section 11.8).

Samples 2 and 3 are initially the same distance below the CSL and they have the same peak stress ratio $\tan \phi'_{m2} = \tan \phi'_{m3}$.

11.8 EQUATIONS FOR PEAK STRENGTH

There are several alternative ways in which engineers represent the peak strength of soil.

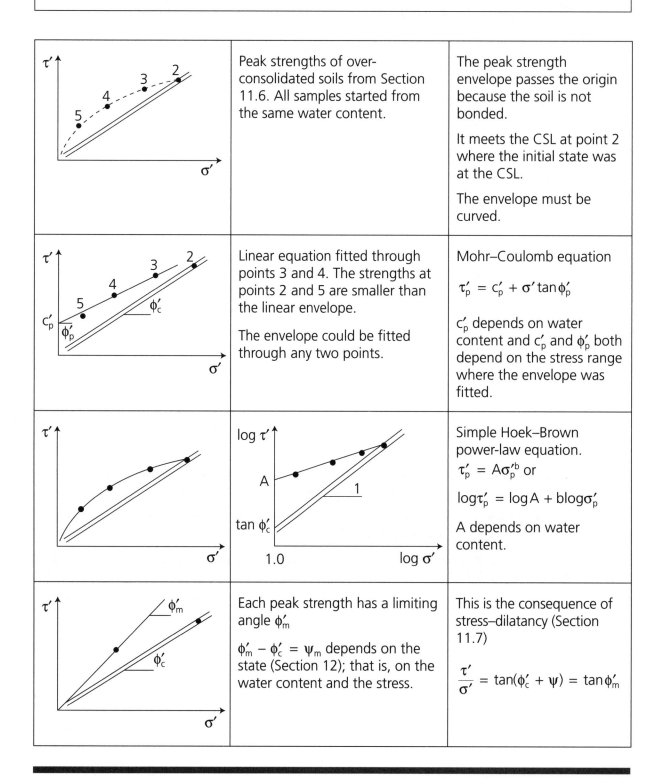

	Peak strengths of over-consolidated soils from Section 11.6. All samples started from the same water content.	The peak strength envelope passes the origin because the soil is not bonded.
		It meets the CSL at point 2 where the initial state was at the CSL.
		The envelope must be curved.
	Linear equation fitted through points 3 and 4. The strengths at points 2 and 5 are smaller than the linear envelope. The envelope could be fitted through any two points.	Mohr–Coulomb equation $\tau'_p = c'_p + \sigma' \tan\phi'_p$ c'_p depends on water content and c' and ϕ'_p both depend on the stress range where the envelope was fitted.
		Simple Hoek–Brown power-law equation. $\tau'_p = A\sigma'^b_p$ or $\log\tau'_p = \log A + b\log\sigma'_p$ A depends on water content.
	Each peak strength has a limiting angle ϕ'_m $\phi'_m - \phi'_c = \psi_m$ depends on the state (Section 12); that is, on the water content and the stress.	This is the consequence of stress–dilatancy (Section 11.7) $\dfrac{\tau'}{\sigma'} = \tan(\phi'_c + \psi) = \tan\phi'_m$

11.9 STRENGTH OF SOIL IN TRIAXIAL COMPRESSION

Soils also reach critical states and peak states in triaxial compression tests. The principles are the same as those for shearing (Sections 11.2 and 11.8) but the parameters differ.

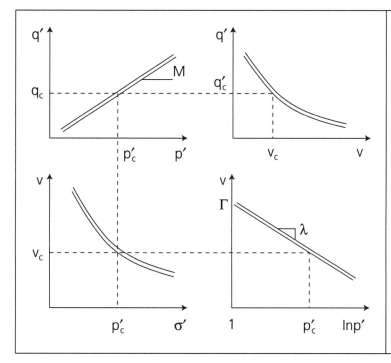

Critical states

$$q'_c = Mp'_c \text{ and } v_c = \Gamma - \lambda \ln p'_c$$

$$q'_c = M\exp\frac{\Gamma - v_c}{\lambda}$$

The critical state parameters M. λ and Γ are material parameters: they depend only on the grains.

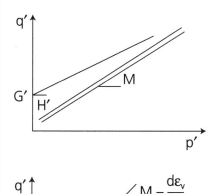

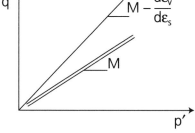

Peak states

Linear envelope:

$$q'_p = G' + H'p'_p$$

Power law:

$$q'_p = \alpha p'^{\beta}_p$$

Stress dilatancy:

$$\left(\frac{q'}{p'}\right)_p = M - \frac{d\varepsilon_v}{d\varepsilon_s}$$

11.10 STRENGTH AND STIFFNESS OF TYPICAL SOILS AND ENGINEERING MATERIALS

These typical values are a guide only. They contrast the relative strengths of typical engineering materials. An important parameter is rigidity which is the ratio of small strain stiffness to strength.

Material	Drained strength	Unconfined compressive strength	Small strain stiffness e_0	Rigidity $= \dfrac{E_0}{UCS}$
Soft clay	$\phi'_m = \phi'_c = 22°$ $\phi'_r = 11°$	50 kPa ($s_u = 25$ kPa)	100 MPa	2000
Stiff clay	$\phi'_m = 30°$ $\phi'_c = 22°$ $\phi'_r = 11°$	300 kPa ($s_u = 150$ kPa)	300 MPa	1000
Loose sand	$\phi'_m = \phi'_c = \phi'_r = 30°$			
Dense sand	$\phi'_m = 40°$ $\phi'_c = 30°$ $\phi'_r = 30°$			
Steel		400 MPa (tensile strength)	200 GPa	500
Concrete		40 MPa	30 GPa	750

Soils are very much weaker and have much lower stiffness than steel and concrete. However, the values of rigidity for soft and stiff soils are comparable and do not differ greatly from those for steel and concrete.

11.11 THINGS TO DO

1. Fill a parallel-sided glass container such as a jam jar about 1/3 full with dry sand. Roll the jar across a table and measure the angle of the continuously failing slope. This is the critical state friction angle (Section 11.2). This is a very good way to measure ϕ'_c of coarse-grained soils.

2. Make a sandcastle from sand with a little water using a plastic cup or a beach bucket. Put the sandcastle on a balance and weigh it. With your hand load the sandcastle from the top and observe the increase in load when it fails. From the unconfined compressive strength and ϕ'_c measured in 1 above calculate the pore pressure in the sandcastle. Dry the sand in an oven or microwave and calculate its water content.

3. Repeat 2 with more and less water and plot a relationship between water content and pore pressure.

4. Repeat 2 and 3 using a suitable fine-grained soil. It may be easier to compact the soil into a tube and push the cylindrical sample out. Make a reasonable assumption for a value for ϕ'_c.

5. In a laboratory do shear tests on loose and dense dry sand. Observe peak and constant volume strengths and volume changes and interpret the observations with any of the equations in Section 11.8.

If you do triaxial tests make sure you do them yourself, including putting the sample into the apparatus, rather than just watching someone else do them.

11.12 FURTHER READING

There is an extensive literature on soil strength in books and papers. Much of this can be confusing and some are even wrong. Make sure that your favourite book deals clearly with:

1. Three strengths, peak, constant volume = critical state and residual. Many books equate the critical state and residual strengths but this is wrong.

2. Undrained strength and effective stress strength.

Consider which of the several equations for peak strength in Section 11.8 best represents soil strength.

11.11 THINGS TO DO

11.12 FURTHER READING

Chapter Twelve

State and State Parameters

In the ground soils and rocks appear to be very different but they are all part of a simple pattern in which behaviour and properties of all geo-materials are related to their state.

As soil is loaded and unloaded isotropically or in 1D it compresses and swells (Sections 9.3 and 10.2). It can reach many combinations of effective stress and water content but all must be inside the normal consolidation line.

Combinations of stress and water content describe the current *state* of the soil. Its behaviour during shear depends on the distance of the current state from its critical states; this distance is a *state parameter*.

The initial state parameter determines how the soil will behave when it is sheared; whether it dilates or compresses in drained shearing or whether pore pressures increase or decrease in undrained shearing.

12.1 STATE: SOILS AND ROCKS

Soils become rocks and rocks become soils as the state changes due to natural geological processes.

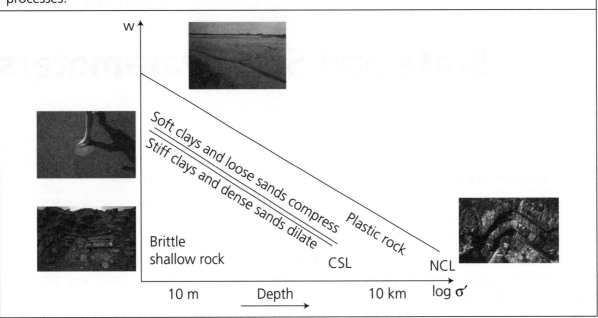

When the rock was folded it was at depths of several kilometers and it was deforming plastically like soft clay. When it is shallow it is brittle. The soft clay has very large water content; when sheared drained it compresses. The dense sand has relatively low water content; when sheared undrained pore pressures decrease and the sand looks dry.

Geological processes change soils to rocks, plastic rocks to brittle rocks and rocks to soils.

State moves by

- Loading from further deposition
- Erosion and re-deposition
- Creep at constant stress
- Engineering construction

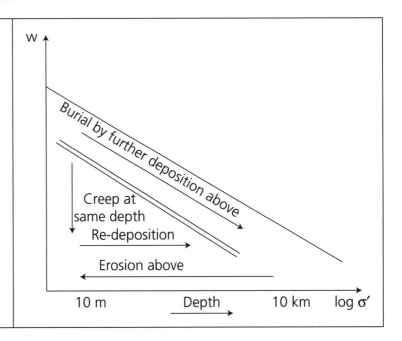

12.2 STATE PARAMETERS

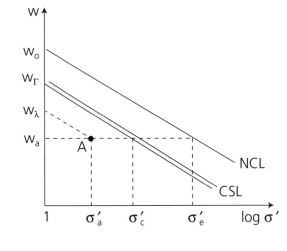

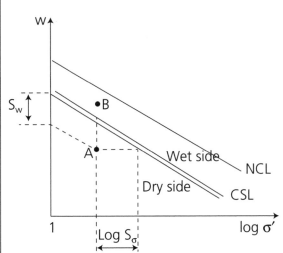

At A the water content is w_a and the stress is σ'_a. The critical stress σ'_c is the stress on the CSL at the same water content (there is also an equivalent stress σ'_e at the NCL).

State parameters S_σ and S_w are

$$S_\sigma = \frac{\sigma'_a}{\sigma'_c}$$

$$S_w = w_\lambda - w_\Gamma$$

These describe how far the state at A is from the CSL. The state at A moves when water content or stress (or both) change and both state parameters change.

When the state is at the CSL

$$S_\sigma = 1 \text{ and } S_w = 0.$$

The region below the CSL containing point A is called the dry side of critical because w_λ is smaller than w_Γ; the region between the CSL and the NCL containing the point B is called the wet side of critical because w_λ is larger than w_Γ; the state of unbonded soil cannot exist outside the NCL.

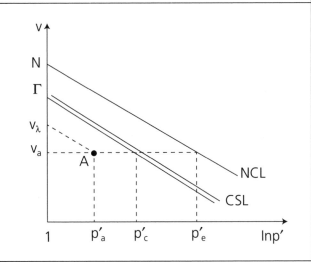

State is also important for triaxial tests. The principles are the same as those for shearing. These state parameters are

$$S_p = \frac{p'_a}{p'_c}$$

$$S_v = v_\lambda - \Gamma$$

12.3 STATE PARAMETERS AND A STATE BOUNDARY

The basic behaviour of a soil is determined by its state and so state parameters can be used to unify all soil behaviour into a single diagram.

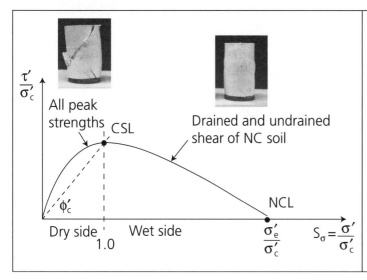

Both τ' and σ' are divided by the critical pressure σ'_c so the horizontal axis is S_σ.

There is a boundary to possible states: this contains the NCL where $\tau' = 0$, the CSL where $\tau'/\sigma' = \tan\phi'_c$ and the peak strengths where $\tau'/\sigma' > \tan\phi'_c$. It also contains a link between the NCL and the CSL where $\tau'/\sigma' < \tan\phi'_c$.

Soils on the dry side form slip planes near the peak strength. Soils on the wet side deform uniformly up to the critical state.

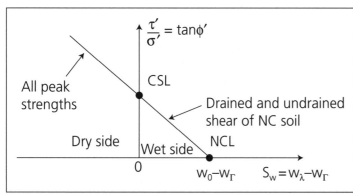

The state boundary can also be shown as the stress ratio τ'/σ' and the state parameter S_w.

The state boundary joins the NCL and the CSL and includes stress ratios larger than $\tan\phi'_c$.

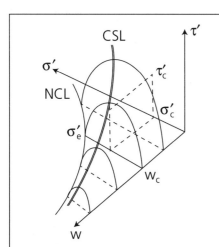

These two diagrams are sections of a 3D graph with axes τ', σ' and w and this graph is a state boundary surface.

Constant volume sections of the surface are a collection of curves with the same shape but different size.

Each curve is associated with a particular value of equivalent stress σ'_e on the NCL and particular values of τ'_c and σ'_c on the CSL.

The state is inside the surface if the soil is unloaded and it cannot be outside the surface.

12.4 STATE PATHS

State diagrams show the complete behaviour for any loading path.

A simple assumption is that soil is isotropic and elastic for states inside the state boundary and yields when the state reaches the state boundary. During shearing of isotropic and elastic soil shear and volumetric effects are decoupled and both S_σ and S_w remain constant.

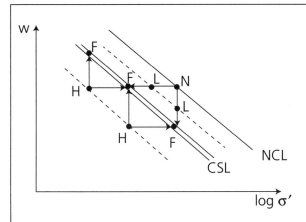

N – normally consolidated; the initial state is on the NCL

L – lightly overconsolidated; the initial state is on the wet side

H – heavily overconsolidated; the initial state is on the dry side.

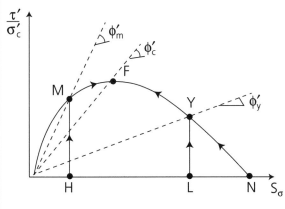

For all initial states and for all drained and undrained shearing the failure is at F where $\phi' = \phi'_c$, $S_\sigma = 1$ and $S_w = 0$

From H the soil yields at M where $\phi' = \phi'_m$. From L the soil yields at Y where $\phi' = \phi'_y$.

Starting from N or L or H the state paths are the same for drained loading with constant pore pressure and undrained loading with constant water content.

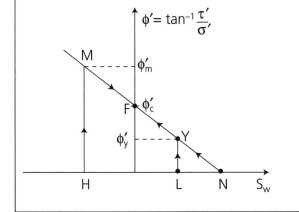

12.5 CREEP AND CHANGE OF STATE

For a soil that does not creep the state can only be changed by loading and unloading. During volumetric creep soil compresses with constant stress and this causes a change of state (Section 10.2).

Coarse-grained soils can be compressed with constant effective stress by vibration. This is a form of creep and has the same consequences for soil state.

At the state at A where the stress is σ'_a

the overconsolidation ratio is $R_0 = \dfrac{\sigma'_m}{\sigma'_a} \geq 1$

and the yield stress ratio is $R_y = \dfrac{\sigma'_y}{\sigma'_a} \geq 1$

Before creep $R_0 = R_y$

After creep the yield stress is larger and $R_0 \neq R_y$.

The yield stress ratio is approximately the same as the state parameter S_σ and is a much better description of soil state than overconsolidation ratio.

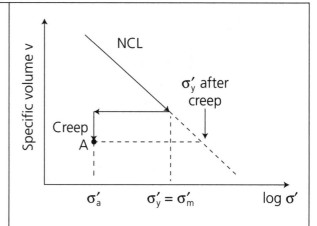

12.6 THINGS TO DO

Study the characteristic behaviour for the nine samples shown in Section 9.4 and the eight samples in Section 11.6. Draw the corresponding state paths using the axes shown in Section 12.4.

Notice that for drained tests with σ' constant the value of σ'_c changes and for undrained tests the value of σ' changes but σ'_c remains constant; in both cases the ratio σ'/σ'_c changes in the same way.

12.7 FURTHER READING

There are very few books that cover the important topic of soil state in a simple and accessible way. Look for diagrams similar to those in Section 12.4, which have accompanying simple explanations.

Chapter Thirteen

Cam Clay and Numerical Modelling

Modern ground engineering practice makes much use of computer-based numerical analyses to investigate both collapse and working load movements. A key part of any numerical analysis is the numerical model for soil behaviour.

A numerical model is simply a set of mathematical equations that represent material behaviour. It need not be exact, only good enough for practical purposes. Hooke's law is a mathematical model for extension and compression of an elastic rod under tensile or compressive stress. Mathematical models that represent the essential features of soil behaviour are more complicated than Hooke's law. They have to take account of the current state and include elasticity, plasticity and drainage.

A reasonably simple mathematical model for soil is called Cam Clay from the river that runs through Cambridge where it was first invented in the 1960s. The basic Cam Clay model has been developed and refined but the original simple form is useful for illustrating the basic features of an elasto-plastic numerical model for soil.

13.1 NUMERICAL MODELS

Principles of numerical modelling

The basic form of a general numerical model is a set of constitutive equations (Section 6.5). These are

$$\{\delta\sigma'\} = [S]\{\delta\varepsilon\} \quad \text{or} \quad \{\delta\varepsilon\} = [C]\{\delta\sigma'\}$$

where $\{\delta\sigma'\}$ contains six stresses $\delta\sigma'_x$, $\delta\sigma'_y$, $\delta\sigma'_z$, $\delta\tau'_{xy}$ $\delta\tau'_{yz}$ $\delta\tau'_{zx}$ and $\{\delta\varepsilon\}$ contains six strains $\delta\varepsilon_x$, $\delta\varepsilon_y$, $\delta\varepsilon_z$, $\delta\gamma_{xy}$ $\delta\gamma_{yz}$ $\delta\gamma_{zx}$. $[S]$ is a stiffness matrix and $[C]$ is a compliance matrix.

The components of the stiffness and compliance matrices should describe the stress–strain behaviour throughout loading and unloading up to and including failure. They contain parameters that are basic soil properties and they probably contain the current stress and water content.

To illustrate the basic features of a numerical model for soil it is simplest to consider the behaviour in isotropic and axisymmetric (triaxial) loading for which the constitutive equations are

$$\delta\varepsilon_s = \frac{1}{3G'}\delta q' + \frac{1}{J_1'}\delta p' \quad \text{and}$$

$$\delta\varepsilon_v = \frac{1}{J_2'}\delta q' + \frac{1}{K'}\delta p'$$

In these the stiffness parameters G', K' and J' vary with the current state and the history of changes of state.

Features of soil behaviour

The principal features of soil behaviour described in Chapter 9 that have a significant influence on the outcome of numerical analyses are:

Drainage: soil strength and stiffness depends on effective stress and pore pressure which depends on drainage. The limiting conditions are fully undrained (constant volume) and fully drained (pore pressures are known).

Strength: soil usually has a peak strength at small strain and a critical state strength at large strain. If the soil is undrained its strength depends on the initial water content and is independent of total stress. If the soil is drained its strength is related to effective normal stress and is frictional and dilatant.

Stiffness: except at very small strain soil is not linear and it is not elastic. When loaded and unloaded there are irrecoverable strains; there are substantial hysteresis loops; stress–strain behaviour is significantly non-linear.

13.2 CAM CLAY

Cam Clay is an elasto-plastic model that combines elasticity, plastic flow, yielding and hardening described in Section 6.8.

The yield curve is obtained by combining stress dilatancy with plastic flow. Elastic behaviour is linear with respect to the log of stress. The model is developed basically from consideration of mechanics and there is very little empirical input.

Yielding

Yielding occurs when the state reaches the yield curve which is similar to that shown in Section 12.4. As the yield stress p'_y increases the yield curve expands.

Elastic straining

For states inside the current yield curve the behaviour is isotropic and non-linear elastic. From Sections 10.4 and 10.8 with the simple approximation $G^{e'} \approx 0.5\ K^{e'}$ corresponding to $v' = 0.2$

$$\delta\varepsilon_v^e = \frac{\kappa}{vp'}\delta p' \quad \text{and} \quad \delta\gamma^e \approx \frac{2\kappa}{3vp'}\delta q'$$

Yield, failure and plastic straining

The flow rule is $dq'/dp' = -d\varepsilon_v^p/d\varepsilon_s^p$, the stress–dilatancy rule is $q'/p' = M - (d\varepsilon_v^p/d\varepsilon_s^p)$ and hence the equation of the yield curve is

$$\frac{dq'}{dp'} = \frac{q'}{p'} - M \quad \text{and} \quad \frac{q'}{Mp'} + \ln\left(\frac{p'}{p'_y}\right) = 0$$

This is a log spiral and each yield curve is associated with a yield stress p'_y. At the NCL $p' = p'_y$ and $q' = 0$; at the CSL $q' = Mp'$ and $\ln(p'_y/p'_c) = 1$ or $p'_y = 2.72p'_c$

Hardening

The hardening law relates expansion of the yield envelope to the plastic volumetric strains.

$$\delta\varepsilon_v = \frac{\lambda}{vp'_y}\delta p'_y \quad \text{and} \quad \delta\varepsilon_v^e = \frac{\kappa}{vp'_y}\delta p'_y$$

and hence

$$\delta\varepsilon_v^p = \frac{\lambda - \kappa}{vp'_y}\delta p'_y$$

13.2 CAM CLAY (CONT)

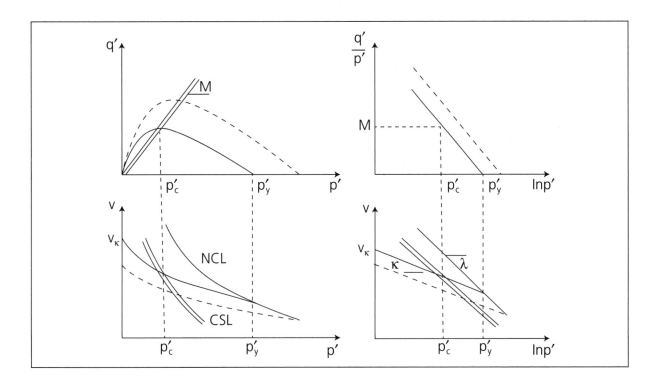

The Cam Clay model is fully explained in several textbooks. The full expression for the state boundary surface is

$$\frac{q'}{Mp'} + \left(\frac{\lambda}{\lambda - \kappa}\right)\ln p' + \left(\frac{\Gamma - v}{\lambda - \kappa}\right) = 1$$

When $q' = Mp'$ this gives $v = \Gamma - \lambda \ln p'$; when $q = 0$ on the NCL $v = N - \lambda \ln p'$ and hence $N - \Gamma = \lambda - \kappa$. The full constitutive equations are

$$\delta\varepsilon_s = \frac{1}{vp'}\left\{\left[\frac{\lambda - \kappa}{M\left(M - \frac{q'}{p'}\right)} + \frac{2\kappa}{3}\right]\delta q + \left[\frac{\lambda - \kappa}{M}\right]\delta p'\right\}$$

$$\delta\varepsilon_v = \frac{1}{vp'}\left\{\left[\frac{\lambda - \kappa}{M}\right]\delta q' + \left[\frac{\lambda - \kappa}{M}\left(M - \frac{q'}{p'}\right) + \kappa\right]\delta p'\right\}$$

During elasto-plastic behaviour when the state is on the state boundary shear and volumetric effects are coupled. Strains depend on basic parameters λ, κ and M and on the current state q', p' and v.

13.3 BEHAVIOUR OF CAM CLAY IN DRAINED LOADING

The paths below are for routine triaxial compression in which total stress $\delta\sigma_r = 0$ and $\delta q = 3\delta p$. For the path 0A the state is below yield for which the behaviour is elastic, and shear and volumetric effects are decoupled. Beyond A the behaviour is elasto-plastic and shear and volumetric effects are coupled.

Wet side of critical – drained loading

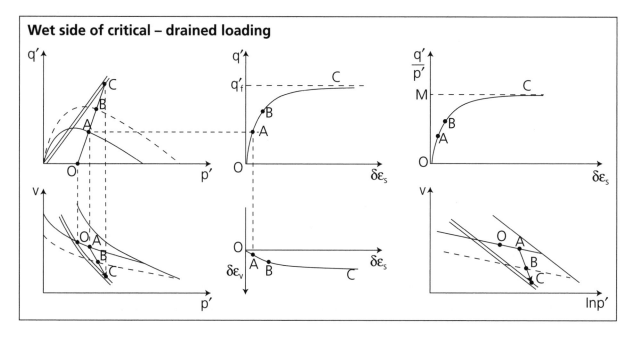

Dry side of critical – drained loading

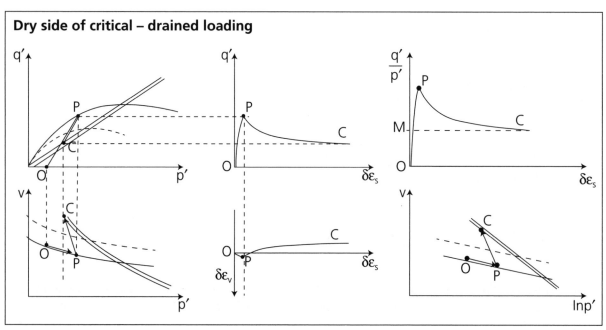

13.4 BEHAVIOUR OF CAM CLAY IN UNDRAINED LOADING

In undrained loading the volumetric strain must remain zero, so $\delta\varepsilon_v = 0$ and this determines the change of pore pressure δu and hence the change of effective stress $\delta p' = \delta p - \delta u$.

Wet side of critical – undrained loading

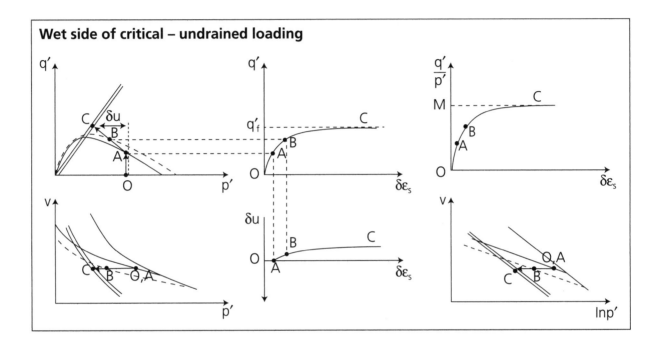

Dry side of critical – undrained loading

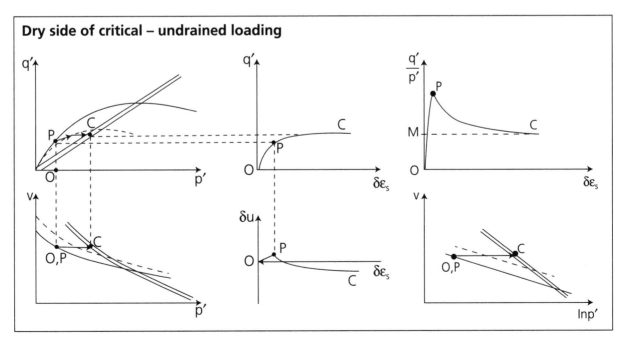

13.5 COMMENTS ON BEHAVIOUR OF CAM CLAY

The behaviours illustrated in Sections 13.3 and 13.4 reproduce many of the key features of soil behaviour shown in Section 9.4.

1. All loading paths reach unique critical states where $q'_c = Mp'_c$ and $v_c = \Gamma - \lambda \ln p'_c$.

2. On the paths OA the states are elastic and shear and volumetric effects are decoupled so during drained loading $\delta \varepsilon_v$ depends on $\delta p'$ and during undrained loading $\delta p' = 0$.

3. Soils on the wet side of critical compress throughout drained loading and pore pressures increase during undrained loading. During drained loading soils on the dry side firstly compress for states below yield and then dilate. During undrained loading the changes of pore pressure are the reverse of the changes of volume.

4. For soils on the wet side of critical the maximum shear stress and the maximum stress ratio both occur at the critical state. For soils on the dry side there are peak states. In drained loading the peak shear stress and the peak stress ratio occur at the same strain. In undrained loading there is no peak shear stress but there is a well-defined peak stress ratio.

There are some features of soil behaviour that are not very well reproduced by the simple Cam Clay model.

1. The change in behaviour from elastic to elasto-plastic as the state reaches the yield curve is abrupt and gives rise to sharp corners on some of the stress–strain curves.

2. The elastic stress–strain curves are not always linear because elastic stiffness parameters vary with vp' but the degree of non-linearity is less than observed in soils.

3. The elasto-plastic behaviour of truly normally consolidated soils becomes indeterminate at very small values of q' near the NCL.

Some of these deficiencies of simple Cam Clay have been addressed and mitigated in developments of the original model.

13.6 FURTHER READING

The original formulation of Cam Clay is in Schofield, A. N. and Wroth, C. P. 1968 *Critical State Soil Mechanics* but this is a very difficult book to read. A simpler derivation is given by Atkinson, J. H. 2007. *The Mechanics of Soils and Foundations*.

Cam Clay is not the only numerical model for soil in current use. Many commercial software packages have specially developed numerical models. Some of these have a sound theoretical basis but others are largely empirical.

It is beyond the scope of this book to examine other numerical models for soil. However, anyone using a numerical model for design of groundworks should examine the basic model used and run independent checks to demonstrate that it gives a reasonable representation of the ground being modelled. The best checks compare numerical analyses of triaxial tests using the proposed model and design parameters against data from high-quality laboratory tests.

13.6 FURTHER READING

Chapter Fourteen

Soil Parameters for Design

Analyses of stability and settlement of geotechnical structures require values for parameters principally for strength, stiffness, permeability and unit weight. These are found from soil descriptions and from in situ and laboratory testing.

Strength and stiffness of concrete depend on the concrete mix, strength and stiffness of steel depend on the steel grade and for both it is relatively easy to select values for design. Soil parameters depend on several variables and these influence the methods required to evaluate them.

14.1 SOIL PARAMETERS

> Values for soil parameters for strength, stiffness etc. depend on the nature of the grains and the grading, on the state, on the strain and on the structure of the soil.

Parameter class	Depend on	Some examples	How to evaluate them
Material parameters	Grading, grain shape, clay mineralogy and plasticity	CS friction angle ϕ_c' NC line C_c and e_o G_s of grains Approx. permeability k Approx. unit weight γ	Correlations with description Tests on samples compressed from a slurry
State-dependent parameters	Soil grains plus the current state of stress and water content	Unit weight γ Peak strength ϕ_m' or c_p' and ϕ_p' Undrained strength s_u Very small strain E_o Permeability k	Relationships between material parameters and state Tests on samples recompressed to the in situ state
State- and strain-dependent parameters	Soil grains plus state plus strain	Stiffnesses E' and E_u 1D compression M' or m_v	Relationships between material parameters, state and strain Tests on samples recompressed to the in situ state and then related to strain
Parameters that depend on structure	All the above plus bonding and layering if present	Parameters for peak strength, undrained strength and compressibility	Best-quality undisturbed samples reconsolidated to the correct state In situ tests

> Reasonable values for many of the important soil parameters can be estimated from soil descriptions and measured in tests on samples that have been disturbed. In some cases, such as measurement of ϕ_c' it is best to test reconstituted and recompressed samples.

14.2 TESTS AND RESULTS

There is a process from basic field and laboratory testing through analysis and interpretation of the results to the parameters that are used in analyses. The first action is to conduct the tests and obtain the basic factual results.

		Basic tests	Results
Description and classification		Grading and plasticity Weigh wet and dry Measure dimensions	Grading curves Atterberg limits Water content Unit weight
Laboratory loading tests		Triaxial compression Shear box 1D consolidation	Plots of stress vs strain Plots of settlement vs time
In situ probing tests		Standard penetration test SPT Cone penetration test CPT	Blowcount N Cone resistance q_c Sleeve resistance f_s
In situ loading tests		Shear vane Pressuremeter	Applied torque Pressure vs expansion
Geophysics		P-wave, S-wave and surface waves	Wave velocities from transmission times
In situ permeability		Pumping	Quantities pumped and surrounding pore pressures

Details of how these tests should be carried out are contained in common national standards and specialised books. Test results are the basic observations that come from each test. Test results are included in the Geotechnical Factual Report (Section 4.3).

14.3 SOURCES OF ERROR IN SOIL TESTS

Basic definition: **Error = true value – reported value**.

All experimental observations contain errors; this is absolutely inevitable and is one of the basic laws of science. There are three basic sources of error:

Systematic error: this is consistent and arises from a feature of the apparatus or the devices used to make the measurement. Examples include friction in the apparatus and between the soil and the apparatus, leakage, approximate calibration constants but there are several others.

Random error: often called scatter or noise, this arises from natural variations between samples that are intended to be the same. Random errors are often quantified as departures from a mean.

Gross error: this is due to a mistake or a major malfunction in the apparatus. These are usually fairly obvious and give unrealistic results; they should be spotted but sometimes are not.

Accuracy and precision

It is important to distinguish between accuracy and precision which have very different meanings.

Accuracy is how far the quoted result is from the true value and it is a measure of the possible error. It should be quoted as a value such as for stress $= \pm$ x kPa; or for force $= \pm$ x kN; or for strain $= \pm$ x %.

Precision is how many significant figures are quoted in the result. It is illogical to quote results with a precision that is much greater than the accuracy.

In ground engineering accuracy and the required precision correspond to about two non-zero figures such as $s_u = 75$ kPa or $\phi' = 27°$ or $\varepsilon_a = 0.12\%$. In calculations this corresponds to the precision that can be achieved using a slide rule. Electronic calculators are capable of giving answers to many significant figures which can give a false indication of the engineering reliability of a result.

14.3 SOURCES OF ERROR IN SOIL TESTS (CONT)

There are some sources of error that are particular to laboratory soil tests.

Actions		Process	Some sources of error
Obtain and prepare samples		Prepare the sample and install it into the apparatus	Sample disturbance. Drainage leads not saturated
Load the sample and make observations		Observe gauges and transducers and convert the readings into force and displacement	Leakages, malfunctioning instruments; incorrect and non-linear calibrations
Process the data		Convert forces and displacements into stress and strain; plot stress–strain curves	Non-uniform strains; friction; mistakes in interpretation

Errors in test results should be evaluated and reported as an absolute value in kPa or per cent strain and not related to a full scale or maximum value.

In routine laboratory tests it is difficult to avoid errors less than ±2–3 kPa and ±0 1% strain and they are often much larger. Factual reports should declare and justify the probable magnitudes of the errors in all measurements but most do not.

14.4 DERIVATION OF GEOTECHNICAL PARAMETERS

The second stage in analysis and interpretation of test results is derivation of geotechnical parameters. In this stage, one parameter or set of parameters is derived from the results of each test. Some parameters are derived through strict definitions and application of theory; other parameters are derived through empirical correlations. Below are some examples of each process.

Tests	Test result	Basic derived parameter
Description and classification	Sieving Weighing Atterberg limits tests	Grading curve Unit weight γ water content w Liquid and plastic limits
Laboratory strength and stiffness tests	Isotropic compression 1D consolidation	K' C_c C_s e_o M' or m_v c_v, k
	Undrained triaxial – u not measured Undrained triaxial – u measured Drained triaxial	s_u, E_u s_u, E_u, ϕ'_c peak strength E', ϕ'_c peak strength, ψ
	Drained shear Undrained shear	ϕ'_c peak strength, ψ s_u
	Drained ring shear Undrained ring shear	ϕ'_r s_{ur}
In situ tests by analysis	Shear vane torque Pressuremeter expansion CPT cone resistance q_c Pumping tests Shear wave velocity	s_u peak and residual s_u and E_u s_u Permeability k G_o
In situ tests by empirical correlations	SPT blowcount N Piezocone; CPT with pore pressure	Strength s_u and ϕ'_m Soil type

The engineer who will use parameters in an analysis should be the one who interprets the test results given in the factual report and derives the parameters. But often this is done by laboratory technicians or field operators.

14.5 CHARACTERISTIC AND DESIGN VALUES

A geotechnical model consists of engineering units that are about the size of a shipping container (Section 4.1). In a typical investigation values for each parameter within each unit will be obtained from various laboratory tests and in situ tests.

A characteristic value is the value of a parameter that represents the properties of an engineering unit and which will be used in analyses of geotechnical structures.

Most commonly used terms avoid precise statistical definitions and are largely self-explanatory.

- Representative value; close to the mean
- Cautious estimate = moderately conservative
- Worst credible

If the design requires failure, such as excavation or flow of grain from a hopper, the worst credible is the largest credible value not the smallest.

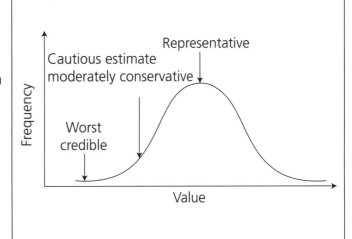

Some parameters, such as undrained strength, vary consistently with depth.

The terms representative and worst credible are also used in this context.

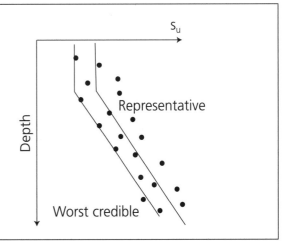

Warning: The meanings of the terms characteristic value and design value are not used consistently in books and codes.

A characteristic value used in a design calculation is sometimes called a 'design value' *before* a factor has been applied. In some codes 'design value' is a characteristic value *after* it has been modified by a partial factor.

14.6 ESTIMATION OF PARAMETERS FROM SOIL DESCRIPTION

Soils are collections of mineral grains packed together loosely or densely. If they are not bonded and are packed randomly (i.e. not layered) their behaviour must depend on the nature of the grains and on the state. It follows that several basic soil parameter values can be estimated from descriptions of the grains and their state.

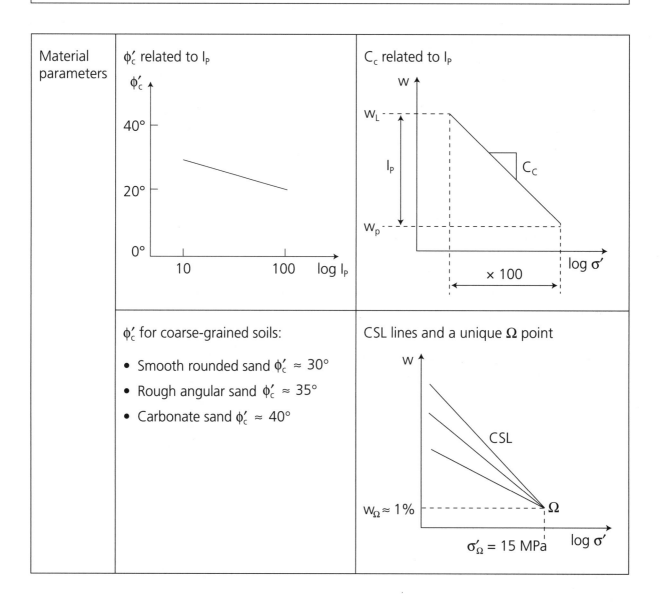

Material parameters	ϕ_c' related to I_P	C_c related to I_P

ϕ_c' for coarse-grained soils:

- Smooth rounded sand $\phi_c' \approx 30°$
- Rough angular sand $\phi_c' \approx 35°$
- Carbonate sand $\phi_c' \approx 40°$

CSL lines and a unique Ω point

$w_\Omega \approx 1\%$

$\sigma_\Omega' = 15$ MPa

14.6 ESTIMATION OF PARAMETERS FROM SOIL DESCRIPTION (CONT)

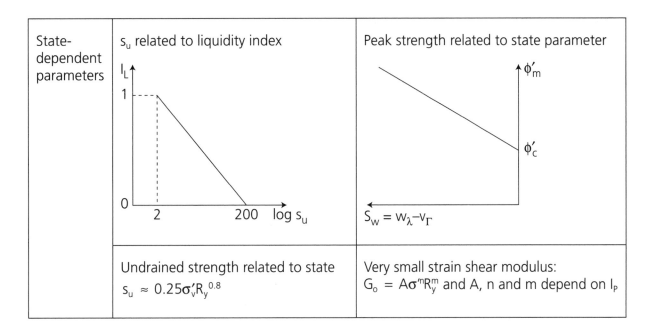

State-dependent parameters	s_u related to liquidity index	Peak strength related to state parameter
	Undrained strength related to state $s_u \approx 0.25\sigma_v'R_y^{0.8}$	Very small strain shear modulus: $G_o = A\sigma'^n R_y^m$ and A, n and m depend on I_P

14.7 SELECTION OF SOIL PARAMETERS FOR DESIGN

Soil parameters for design calculations have to be chosen by the designer from a combination of

- estimation from soil description
- laboratory tests
- in situ tests

The value that is selected for design depends on:

- whether the analysis is to prevent collapse or limit movement;
- what factors will be applied in the analyses (Section 20.2);
- which parameter is being considered (e.g. peak, CS or residual strength).

Selection of characteristic or design values is mostly a matter of skill and judgment. There is rarely sufficient data to justify formal statistical analysis.

It would be unwise to design geotechnical structures solely on the basis of parameters estimated from descriptions. But if the estimated values differ substantially from those measured in laboratory and in situ tests the reason needs to be investigated.

The differences may be due either to errors in the tests or because the soil has some special features that modify its behaviour. In either case the differences should be reconciled.

14.8 FURTHER READING

The laboratory and in situ tests that are used in practice to measure soil parameters are often determined by national codes and specifications. These generally give details of how the tests should be conducted and how the data should be recorded. Some give details of how the test data should be converted into geotechnical parameters.

Most codes do not give precise advice on which of the several alternative strength parameters (such as peak, critical state or residual) should be used for design nor do they give clear advice on how stiffness parameters should be selected from highly non-linear stress–strain curves. Few codes give any helpful advice on assessing errors or validating data.

There are textbooks that give details of laboratory and in situ testing relevant to different national standards and codes. Again few of these give clear advice on choice of parameters, assessment of errors and validation of data.

Part D

Geotechnical Engineering

Chapter Fifteen

Routine Analyses of Geotechnical Structures

Analyses should achieve designs that prevent failure but failure can mean several different things. It can mean failure to prevent a collapse or it can mean failure to meet requirements of movement and leakage that do not themselves cause a collapse. Different geotechnical analyses investigate collapse (often called an ultimate limit state) and performance, particularly ground movements (often called a serviceability limit state).

The conditions that govern the behaviour of a geotechnical structure are basically the same as those for any structure; forces and stresses must be in equilibrium, displacements and strains must be compatible and stresses and strains must be linked through material properties (Section 5.2).

For many geotechnical structures it is difficult to satisfy all these requirements simultaneously and, most routine analyses make simplifying assumptions. These lead to a number of standard methods.

15.1 BASIC ANALYSES

Requirements		Topic	Routine analyses
Prevent collapse of slopes and foundations		Slope stability Bearing capacity Ultimate limit state	Limit equilibrium Plastic bounds Standard charts
Limit movements of foundations		Allowable movement at working loads Serviceability limit state	Elastic analyses 1D compression
Determine rates of settlement of foundations		Consolidation time	1D consolidation
Prevent collapse of retaining walls		Earth pressures Overall stability Loads in props Stress in the wall	Rankine and Coulomb Limit equilibrium
Assess drainage		Rate of flow Pore pressures	Flownet

15.2 LIMIT EQUILIBRIUM ANALYSES

This is the common method to examine collapse of geotechnical structures. The process is in three steps:

1. Choose a set of slip surfaces that forms a mechanism that permits large movement.

2. Show that blocks in the mechanism are in equilibrium with the soil strength on the slip surfaces.

3. Examine all possible mechanisms and find the one that gives the most critical solution.

Slip surface	Mechanism	Equilibrium calculation	Undrained soil $\tau = s_u$	Drained soil $\tau' = \sigma' \tan\phi'$
Planes		Draw a closed force polygon		
Circular arc		Equilibrium of moments	$Wx = s_u LR$	$T = T'$ depends on N' so this case cannot be solved by moments without simplifying assumptions.
General shapes and slices		Sum equilibrium of all slices		The slice is statically indeterminate and a solution requires assumptions. The simple (Swedish) solution assumes $E = 0$. The Bishop method assumes $\theta = 0$.

The main difficulty is to find the mechanism that gives the most critical solution. If this is not found by the designer it will be found by a collapsing structure.

Many of the tables and charts given in texts for routine analyses of collapse of geotechnical structures were derived using limit equilibrium analyses.

15.3 UPPER AND LOWER BOUND PLASTICITY METHODS

These solutions are theoretically correct for a material that is perfectly plastic. This means that the shear stress remains constant and the angles of dilation and friction are the same ($\psi = \phi'$). The processes are:

Upper bound: choose any compatible mechanism of slip surfaces and apply a small displacement. If the work done by the internal stresses (E) equals the work done by the external forces failure must occur; the external loads are unsafe. Work is the product of force and displacement: for undrained loading $E = s_u L \delta$ and for drained loading with $\phi' = \psi$ then $E = 0$.

Lower bound: for a set of internal stresses that do not exceed the soil strength and which are in equilibrium with the external loads collapse cannot occur; the external loads are safe.

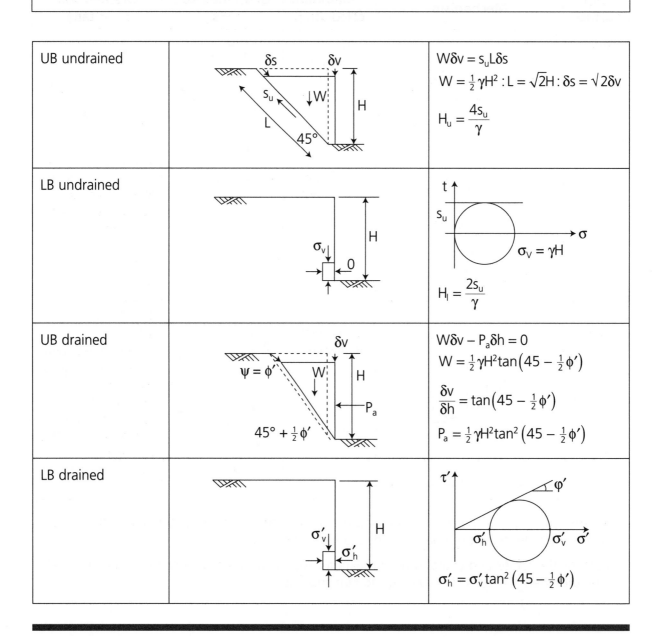

UB undrained		$W\delta v = s_u L \delta s$ $W = \frac{1}{2}\gamma H^2 : L = \sqrt{2}H : \delta s = \sqrt{2}\delta v$ $H_u = \dfrac{4s_u}{\gamma}$
LB undrained		$\sigma_v = \gamma H$ $H_l = \dfrac{2s_u}{\gamma}$
UB drained		$W\delta v - P_a \delta h = 0$ $W = \frac{1}{2}\gamma H^2 \tan\left(45 - \frac{1}{2}\phi'\right)$ $\dfrac{\delta v}{\delta h} = \tan\left(45 - \frac{1}{2}\phi'\right)$ $P_a = \frac{1}{2}\gamma H^2 \tan^2\left(45 - \frac{1}{2}\phi'\right)$
LB drained		$\sigma_h' = \sigma_v' \tan^2\left(45 - \frac{1}{2}\phi'\right)$

15.4 FOUNDATIONS ON ELASTIC SOIL

There are standard solutions for stresses and settlements below circular or square shallow foundations on soil that is assumed to be elastic.

The basic solutions are usually given in terms of influence factors for stress I_σ and for settlement I_ρ.

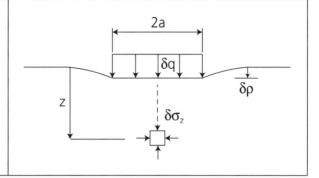

Stress	Undrained settlement	Drained settlement
$\delta\sigma_z = \delta q I_\sigma$ σ_z is independent of E and ν	$\delta\rho_u = \delta q \cdot 2a \cdot \dfrac{1 - \nu_u^2}{E_u} \cdot I_\rho$ and, with $\nu_u = 0.5$ $\delta\rho_u = \delta q \cdot 1 \cdot \dfrac{5a}{E_u} \cdot I_\rho$	$\delta\rho_d = \delta q' \cdot 2a \cdot \dfrac{1 - \nu'^2}{E'} I_\rho$

Values of I_σ and I_ρ are 1.0 immediately below the foundation and decrease rapidly with depth. Most books on ground engineering give values for circular, square and irregularly shaped foundations.

For circular or square foundations on fine-grained soil there will be undrained settlements $\delta\rho_u$ followed by further settlements due to consolidation $\delta\rho_c$. For isotropic and elastic soil:

$$G' = \frac{E'}{2(1 + \nu')} = G_u = \frac{E_u}{2(1 + \nu_u)}$$

$$\frac{\delta\rho_u}{\delta\rho_d} = \frac{1}{2(1 - \nu')} \quad \text{and} \quad \delta\rho_c = \delta\rho_d - \delta\rho_u$$

For long and wide foundations on relatively thin beds of soil the deformations are one dimensional and

$$\delta\rho_u = 0$$

$$\delta\rho_c = \delta\rho_d = \frac{1}{M'} D\delta\sigma' = m_v D\delta\sigma'$$

where D is the depth of compressing soil.

The greatest source of error in calculation of foundation settlement is in selection of an appropriate value for soil stiffness E' or E_u or M' from non-linear stress–strain behaviour (Section 10.6).

15.5 STEADY SEEPAGE

The basic analysis for steady-state seepage is given in Section 7.4. This depends on the construction of a 'square' flownet consisting of flowlines and equipotentials that have to satisfy internal geometry and boundary conditions.

Simple flownet for seepage between a river and a canal.

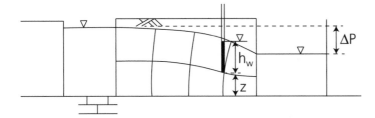

Rate of seepage (m³/s through a slice 1m thick out of the page)

$$\frac{\delta Q}{\delta t} = k \frac{N_f}{N_d} \Delta P$$

On any equipotential the potential is

$$P = z + h_w = z + \frac{u}{\gamma_w}.$$

At the top flowline $u = 0$ and δP is the same between any two equipotentials.

Examples of typical flownets.

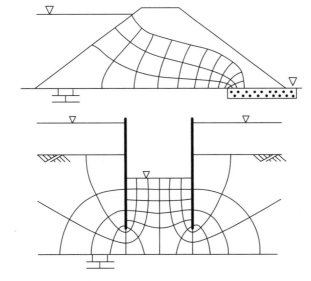

15.6 CONSOLIDATION SETTLEMENT

The basic processes of consolidation as excess pore pressures dissipate at constant total stress and cause changes of effective stress and volumetric strain are in Section 10.9. These apply to consolidation in the ground below embankments and wide foundations.

Magnitude of the final settlement

This depends on $\Delta\sigma_v$ and the 1D stiffness M' or compressibility m_v

The final settlement, when $\bar{u} = 0$ and $\Delta\sigma' = \Delta\sigma$ is

$$\Delta\rho_\infty = m_v D\Delta\sigma' = \frac{1}{M'} D\Delta\sigma'$$

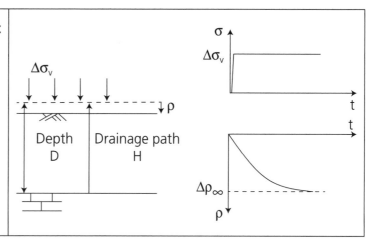

Time for consolidation

The degree of consolidation and the time factor are

$$U_t = \frac{\Delta\rho_t}{\Delta\rho_\infty} \quad \text{and} \quad T_v = \frac{c_v t}{H^2}$$

When $U_t = 1$ then $T_v \approx 1$ and $t_\infty \approx \frac{H^2}{c_v}$

For $U_t < 0.5$ the degree of consolidation is approximately

$$U_t = \frac{2}{\sqrt{\pi}}\sqrt{T_v}$$

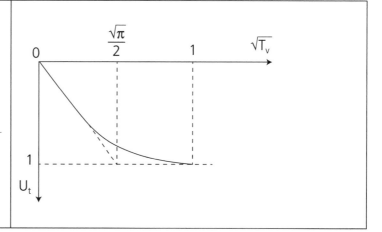

Drainage path length

The drainage path H is the distance travelled by a drop of water escaping to a drain where $\bar{u} = 0$.

This depends on the configuration of drainage layers or installed drains.

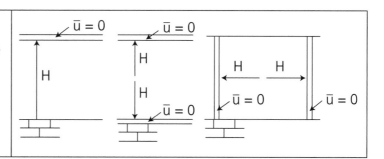

15.7 FURTHER READING

Most textbooks describe these basic analyses in much greater detail than here and most give derivations.

Nearly all describe the limit equilibrium method in Section 15.2. Some give the theory of the upper and lower bound plasticity methods in Section 15.3.

There are books that give solutions and design charts specifically for elastic soils similar to those in Section 15.4. These include solution charts for a very wide variety of foundation shapes and depths, for piles and retaining walls and for layered and anisotropic elastic soils.

Most standard textbooks derive the basic equations for 1D consolidation and give mathematical solutions. They also derive the basic flownet analyses for 2D steady-state seepage. There are books that give examples of 2D flownets for a variety of cases.

Chapter Sixteen

Stability of Slopes

Slopes may be formed by natural erosion of a sea cliff or by a river in a valley or during construction works by excavation into the ground or by compacting fill into embankments. If a slope is too steep or too high it will slip and there will be a landslide.

The usual criterion for design of a slope, or assessment of an existing slope, is to ensure that there is an adequate margin of safety against landsliding. A slope is said to have failed when there are very large movements; it is not usual to calculate movements separately.

16.1 SLOPE FAILURES

Slopes can remain stable because the ground can mobilise a shear stress; water cannot have a slope because it has no shear strength.

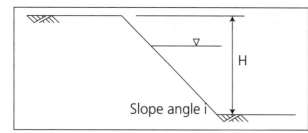

The variables in a failing slope are

- Critical height H_c and critical angle i_c
- Soil strength s_u or ϕ'
- Pore pressures u
- Free water pressures p_w

Soil slope failures generate new slip surfaces.

Shallow slips: slip plane parallel with the slope.		
Deep-seated slips: slip plane circular or a general shape.		
Embankment on soft ground; similar to bearing capacity of a foundation.		

Rock slope failures are governed principally by the orientations of preexisting joints.

Sliding on preexisting slip plane		
Toppling; geometric instability.		

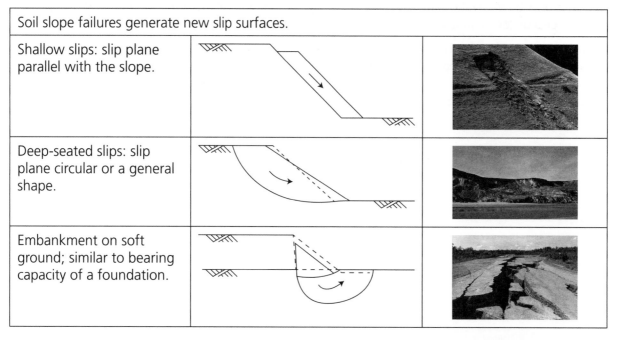

16.2 STRESS CHANGES IN SLOPES

As a slope is made steeper or deeper the average total stress in the ground decreases and the average shear stress increases: the ratio τ/σ is approximately the same as the slope angle tan i. The examples below are for isotropic and elastic soil so if the soil is undrained σ' remains constant. The final steady state pore pressure is taken to be the same as the initial pore pressure.

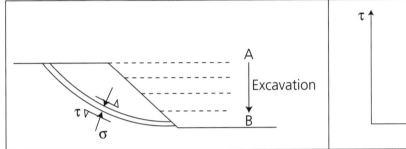

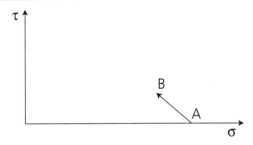

Undrained failure	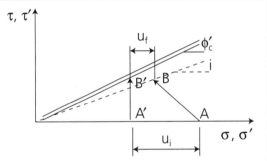
During undrained excavation A to B pore pressures decrease (and may become negative).	
The slope fails when the effective stress reaches the CSL.	

Drained failure	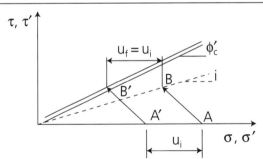
As the slope is excavated A to B pore pressures remain constant.	
The slope fails when the effective stress reaches the CSL.	
At failure $i = \frac{1}{2}\phi'_c$	

Undrained excavation and consolidation	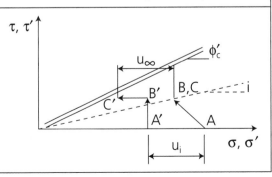
After undrained excavation $\bar{u}$ is negative ($u < u_\infty$). As the soil consolidates u increases and effective stresses reduce with time. The effective stress C′ may or may not reach the CSL.	
During consolidation the water content increases and the soil becomes weaker.	

16.3 BASIC SLOPE STABILITY SOLUTIONS

There are simple standard tables and charts for slope stability. These can be used for preliminary designs and should be used to check computer-based analyses.

Remember that if the soil is undrained pore pressures decrease during excavation and with time after excavation pore pressures increase, water content increases, effective stress decreases and soil gets weaker; generally the stability of a slope becomes worse with time.

Undrained vertical cuts and steep slopes $H_c = \dfrac{N_s s_u}{\gamma}$	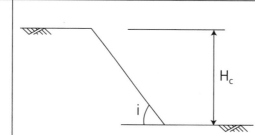 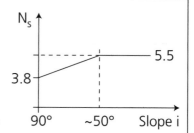
Drained slopes with shallow slips Dry soil: $i = \phi_c'$ Steady-state seepage $i \approx \frac{1}{2}\phi_c'$	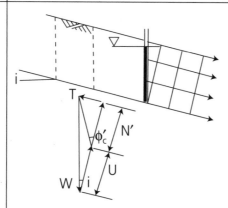
Drained slopes with deep circular slips $F_s = m - n r_u$ $r_u = \dfrac{u}{\sigma_v}$	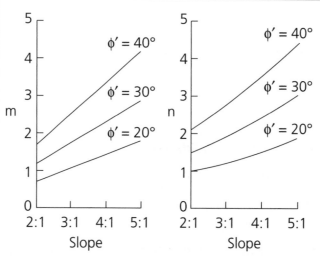 (Bishop, A. W. and N. R. Morgenstern. 1960. Stability coefficients for earth slopes, *Geotechnique*, 10, 129–150.)

16.4 PORE PRESSURES IN SLOPES

Pore pressures have a major influence on slope stability through their contributions to effective stress and strength.

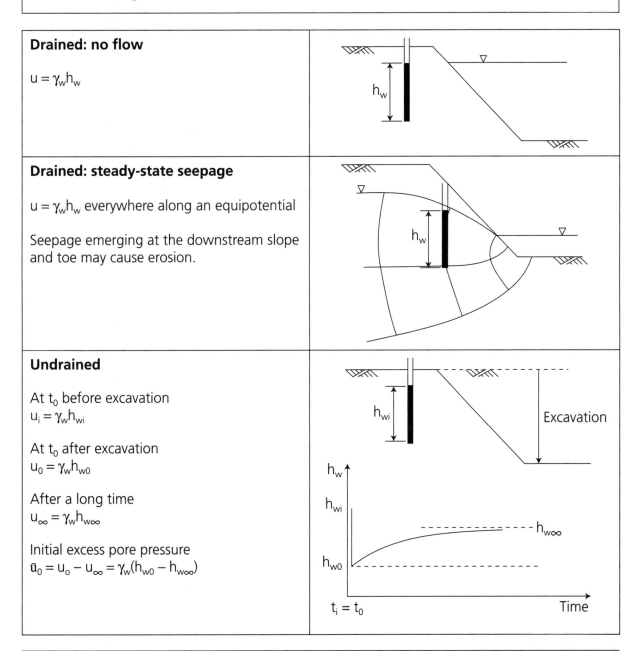

Drained: no flow	
$u = \gamma_w h_w$	

Drained: steady-state seepage

$u = \gamma_w h_w$ everywhere along an equipotential

Seepage emerging at the downstream slope and toe may cause erosion.

Undrained

At t_0 before excavation
$u_i = \gamma_w h_{wi}$

At t_0 after excavation
$u_0 = \gamma_w h_{w0}$

After a long time
$u_\infty = \gamma_w h_{w\infty}$

Initial excess pore pressure
$\bar{u}_0 = u_o - u_\infty = \gamma_w(h_{w0} - h_{w\infty})$

Safety warning: If pore pressures in a slope have not reached a steady state they will almost certainly be increasing with time and the slope is gradually becoming less stable. Just because a slope is stable now does not mean it will be stable in 5 min time. Too many people have been killed by collapsing trenches and small excavations.

16.5 A HOLE IN THE BEACH

A simple hole dug in the beach illustrates many of the basic features of slope stability.

Safety warning: If you try this experiment by digging a deep hole remember that the vertical face above the water table relies on suctions in the pore water and it is potentially unstable.

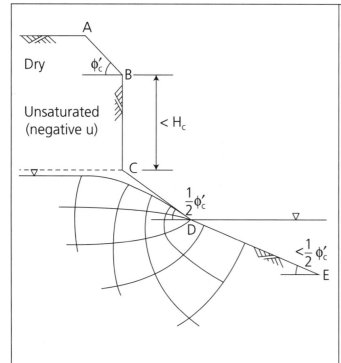

Dry soil $i = \phi'_c$

Soil above the water table may be saturated or unsaturated and its strength s_u depends on suction which depends on grain size (Sections 7.2 and 7.3).

$$H_c = \frac{N_s s_u}{\gamma}$$

In the soil between the water table and free water in the excavation seepage is approximately parallel with the slope

$$i \approx \tfrac{1}{2}\phi'_c$$

Below free water level there is upward seepage

$$i < \tfrac{1}{2}\phi'_c$$

16.6 THINGS TO DO

Most people have already dug a hole in the beach like that in Section 16.5 when they were children but without observing the soil mechanics principles. It is instructive to do it again and relate the observations to the principles.

Investigate the influence of water on slope stability. (These experiments can be done at home; use clean sand about 1 mm size.)

1. Slowly pour dry sand into a conical heap; the slope angle is the critical state friction angle ϕ'_c.

2. Very slowly pour the sand totally underwater in a bowl being careful to avoid eddy currents in the water. The slope angle is the same as for dry sand. (Water does not lubricate soil.)

3. Rapidly drain or syphon the water from the bowl again being careful to avoid eddy currents. When the bowl is empty water drains from the sand towards the slope which then becomes shallower due to seepage.

4. Make a small dam like that shown in Section 7.6. The steady-state seepage through the dam is like that shown in Section 16.4. Observe erosion at the toe which will eventually lead to failure like that shown in Section 1.4.

16.7 FURTHER READING

Slope stability is included in most books on soil and rock mechanics and there are several specialist books on slope engineering.

In Section 16.3 there is a simple chart of N_s for undrained stability and simple charts of m and n for drained stability. There are similar charts for other slope geometries and ground conditions in books and publications.

There are several commercial software packages for slope stability. Most are based on limit equilibrium analyses and make various assumptions to avoid the statical indeterminacy problem in Section 15.2.

Before using a software slope stability package it is essential to carry out some simple analyses with pencil and paper using the analyses in Section 16.3 or similar analyses.

Chapter Seventeen

Foundations

A foundation transfers load into the ground. A foot or a car tyre is a simple foundation. As the loading on a foundation increases it settles and it may rotate. A small building with a deep basement unloads the ground which then swells and the foundation heaves.

If the settlement becomes large the foundation is said to have reached its bearing capacity although it may not have collapsed because more load can be added and there will be further settlements.

Only the simple case of a foundation with a central vertical load will be considered here. Other cases of inclined and eccentric loading are analysed in specialist textbooks on foundation engineering.

17.1 FOUNDATION TYPES

Foundations on relatively stiff and strong ground can be shallow. If the loading would cause failure or large settlement the base of the foundation would be deeper where the ground is stronger and stiffer.

Shallow foundations and footings

D small compared to B and L May be circular, or square or rectangular Bearing pressure $q = \dfrac{F + W}{BL}$ It is best to consider the applied load F and the weight W of the foundation separately		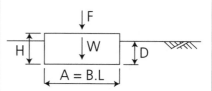
Embankment on soft ground Usually long out of the page H small compared to B Bearing pressure $q = \gamma H$		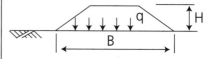

17.1 FOUNDATION TYPES (CONT)

Pile foundations

Driven piles

The resistance R can be estimated from the penetration s for each hammer blow from

$Rs = \eta Wx$:

η = efficiency which depends on type of hammer

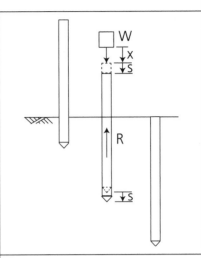

Bored and cast in situ piles

Bored hole supported by slurry and then replaced by poured concrete. Resistance from base Q_b and from shaft Q_s

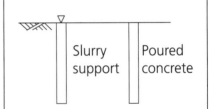

Deep foundations and pile groups

For a deep foundation the depth is 1–3 times the width. Resistance to load F is derived from the base Q_b and from sides Q_s

When the piles in a group are relatively closely spaced the group acts as a deep foundation

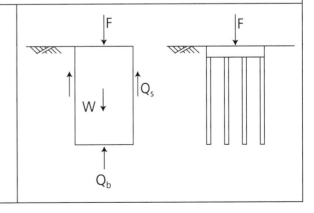

17.2 STRESS CHANGES BELOW A SHALLOW FOUNDATION

As the load on a foundation increases both the average total normal stress and the average shear stress in the soil below the foundation increase. Pore pressure in the ground depends on whether the soil is drained or undrained or consolidating.

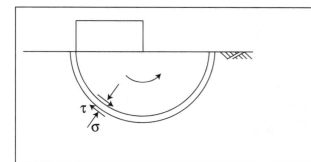

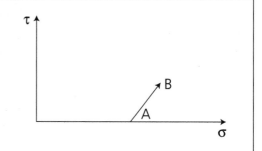

Undrained loading of a stable foundation

During undrained loading of isotropic elastic soil both water content and σ' remain constant and the stress path is A' to B'.

The effective stresses do not reach the CSL so the foundation does not collapse.

The excess pore pressures are $\bar{u}$ and the undrained settlement is $\Delta\rho_u$.

As the soil consolidates excess pore pressures dissipate, the effective stress increases and the water content decreases. The final pore pressure u_∞ is the same as the initial pore pressure u_i. Consolidation settlements are $\Delta\rho_c$.

The effective stress moves away from the failure line along B' to C' so the foundation becomes safer with consolidation time.

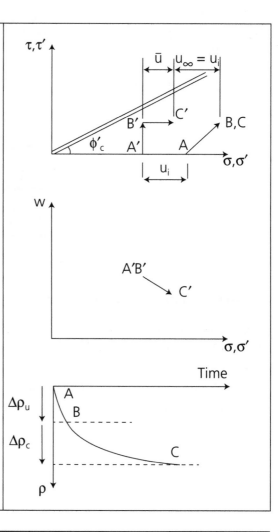

17.2 STRESS CHANGES BELOW A SHALLOW FOUNDATION (CONT)

Undrained failure of a foundation

During undrained loading of isotropic elastic soil both water content and σ' remain constant and the stress path is A' to B'.

The calculated collapse load is when the stress path reaches the failure line at B'.

The pore pressure at collapse u_f is larger than the initial pore pressure u_i.

As the load approaches the calculated collapse load settlements increase quickly as a slip plane develops.

Collapse is at some arbitrary point B where the calculated bearing capacity is q_c.

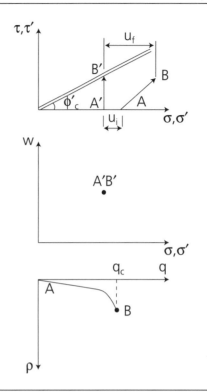

Drained failure of a foundation

During drained loading pore pressures remain constant.

As the effective stress increases the water content decreases; the foundation settles and the soil becomes stronger.

As the stress path approaches the failure line at B' settlements increase slowly as a slip plane develops.

Collapse is at some arbitrary point B where the calculated bearing capacity is q_c.

Further load can be added. As the water content decreases and the effective stress increases the soil becomes stronger as the foundation settles.

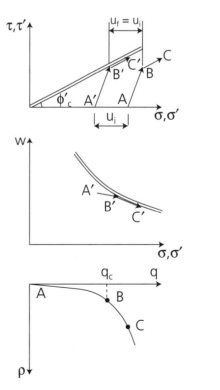

17.3 LOADING AND MOVEMENT OF A FOUNDATION

In a general case there are structural loads F and water pressures at the top p_{wt}, the foundation has a self-weight W, water pressures u_b and soil pressures q' act at the base and there are shear stresses τ_s on the sides. The vertical forces on any foundation must be in equilibrium.

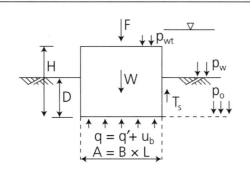

Loads $= F + W + p_{wt}A$

Resistances $= (q' + u_b)A + \tau_s D(2B + 2L)$

For equilibrium: loads = resistances

Shallow foundation: D is small and $\tau_s = 0$
Building with a basement W = 0
Usually ground is not submerged so $p_{wt} = p_w = 0$

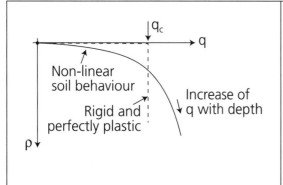

The bearing capacity of a foundation is the stress that causes it to collapse.

For a typical foundation the load–settlement response is a smooth curve and there is no clearly defined collapse.

Routine analyses assume that the soil is rigid and then perfectly plastic so there is a well-defined *calculated* bearing capacity q_c.

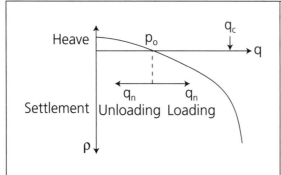

Settlements and heave are caused by changes of stress which are *net* pressures.

Gross total bearing pressure = q = total stress between soil and the base of the foundation.

Net bearing pressure = change of total stress
$\Delta q = q_n = q - p_o$

Gross effective bearing pressure $= q' = q - u_b$

Net effective bearing pressure = change of effective stress $\Delta q' = q'_n = q' - p'_o$

17.4 FLOATING FOUNDATIONS

There are two slightly different cases of a floating foundation.

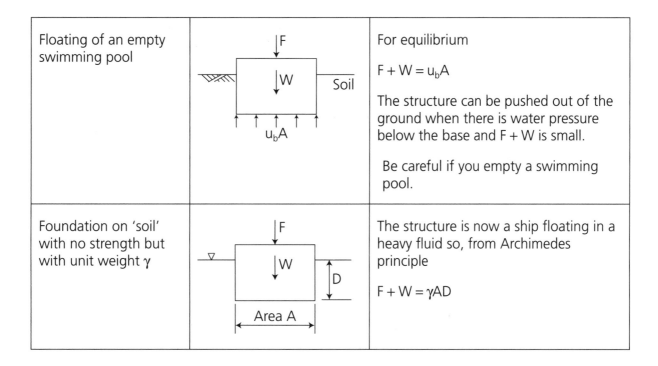

Floating of an empty swimming pool		For equilibrium
		$F + W = u_b A$
		The structure can be pushed out of the ground when there is water pressure below the base and $F + W$ is small.
		Be careful if you empty a swimming pool.
Foundation on 'soil' with no strength but with unit weight γ		The structure is now a ship floating in a heavy fluid so, from Archimedes principle
		$F + W = \gamma AD$

A good test of a bearing capacity equation is to put $s_u = 0$ or $c' = \phi' = 0$ into the bearing capacity factors and the equation should reduce to $F + W = \gamma AD$ which is Archimedes principle. It is a good idea to check that the equations you are using meet this test.

17.5 BEARING CAPACITY FACTORS

There are standard solutions for simple cases given as bearing capacity factors. The basic bearing capacity equations below use parameters defined in Section 17.3 and satisfy vertical equilibrium.

These analyses and factors are for vertical and central loadings: textbooks and codes give additional factors that take account of inclined and eccentric loading and foundations with a sloping base.

Loading	Basic bearing capacity equations	Bearing capacity factors
Undrained loading	$q_c - p_0 = s_u N_c$ N_c = bearing capacity factor that depends on shape and depth. For a long foundation with $D = 0$ the exact solution is $N_c = (2 + \pi) = 5.14$ The increase of N_c with depth is due to shearing stress in soil above the base of the foundation.	 (From Skempton, A. W. 1951. 'The bearing capacity of clays', *Proceedings of Building Research Congress*, Vol. 1, pp. 180–189, ICE, London.)

The chart shows N_c (vertical axis, 0 to 10) versus D/B (horizontal axis, 0 to 6), with curves for "Circular or square" and "Long (L >> B)", and the relation $\dfrac{N_c \text{ (rectangular)}}{N_c \text{ (square)}} = 0.84 + 0.16 \dfrac{B}{L}$

17.5 BEARING CAPACITY FACTORS (CONT)

Loading	Basic bearing capacity equations	Bearing capacity factors
Drained loading	$q'_c - p'_0 = c'N_c + \frac{1}{2}B(\gamma - \gamma_w)N_\gamma$ $+ (N_q - 1)p'_0$ N_q = bearing capacity factor for contribution of soil above the base of the foundation For a long foundation the exact solution is $N_q = \tan^2\left(45 + \frac{1}{2}\phi'\right)e^{\pi\tan\phi'}$ N_c = bearing capacity factor for contribution of effective stress cohesion where $N_c = \dfrac{(N_q - 1)}{\tan\phi'}$ N_γ = bearing capacity factor for contribution of soil below the base of the foundation. An approximate solution is $N_\gamma = 2(N_q - 1)\tan\phi'$	

17.6 SETTLEMENT OF SHALLOW FOUNDATIONS

For simple settlement analyses, wide foundations, usually road, rail and small water retaining embankments are assumed to be 1D and stiffness M' found from 1D consolidation tests. For foundations that are not 1D settlements are calculated using standard elastic solutions with Young's modulus found from undrained or drained triaxial tests.

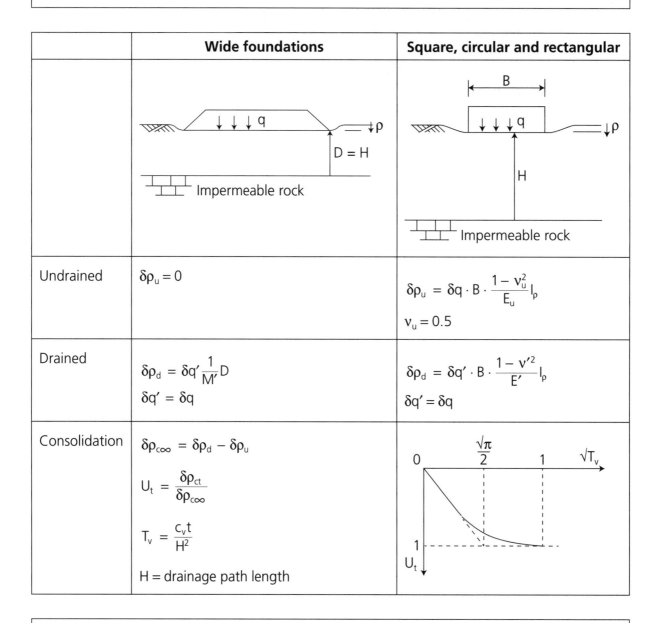

	Wide foundations	**Square, circular and rectangular**
Undrained	$\delta\rho_u = 0$	$\delta\rho_u = \delta q \cdot B \cdot \dfrac{1-\nu_u^2}{E_u} I_\rho$ $\nu_u = 0.5$
Drained	$\delta\rho_d = \delta q' \dfrac{1}{M'} D$ $\delta q' = \delta q$	$\delta\rho_d = \delta q' \cdot B \cdot \dfrac{1-\nu'^2}{E'} I_\rho$ $\delta q' = \delta q$
Consolidation	$\delta\rho_{c\infty} = \delta\rho_d - \delta\rho_u$ $U_t = \dfrac{\delta\rho_{ct}}{\delta\rho_{c\infty}}$ $T_v = \dfrac{c_v t}{H^2}$ H = drainage path length	

Soil stiffness is non-linear and depends on the current state of stress and water content. Laboratory tests to measure the stiffness parameters M' or E' or E_u should represent the average stresses and strains expected in the ground.

17.7 SINGLE PILES

Piles transfer loads to lower levels where the ground is stronger and stiffer. Their capacity arises from a combination of end bearing and side shear stress.

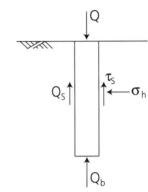

For equilibrium

$$Q = Q_b + Q_s$$

In an end bearing pile most resistance is derived from the base:

$$Q_b = q_b A_b$$

In a friction pile most resistance is derived from the shear stress on the shaft:

$$Q_s = \tau_s A_s$$

For practical purposes the unit weights of soil and concrete are the same so the weight of the pile can be ignored.

Capacity of single piles		
	Undrained loading	**Drained loading**
End resistance	$Q_b = q_b A_b$	$Q_b = (q'_b + u)A_b$
	$q_b = s_u N_c$	$q'_b = (\gamma - \gamma_w)DN_\gamma$
	$N_c \approx 9$	N_γ depends on ϕ'
Side resistance	$Q_s = \tau_s A$	$Q_s = \tau'_s A$
	$\tau_s = \beta s_u$	$\tau'_s = \sigma'_h \tan\delta' = \alpha \sigma'_v \tan\phi'$
	$\beta \approx 0.5$	$\alpha \approx 0.5$

17.8 THINGS TO DO

It is illustrative to do some experiments loading model foundations on different soils. At home you can do these with kitchen materials and utensils or you can do them in a laboratory.

Find some coarse-grained material (sugar or dry sand) and some warm butter and cold butter. Estimate the friction angle of the sugar or sand by pouring it into a heap. Estimate the undrained strength of the butter as though it were a fine-grained soil. For a shallow foundation, use coins of different sizes. For a pile use a pencil.

Push foundations into different materials and observe how the load varies with settlement or displacement. The load can be measured by placing the 'ground' on kitchen or bathroom scales and observing the increase in load as you push the foundation into the soil.

Alternatively, and better, do these experiments in a soils laboratory using a triaxial loading frame and instruments to measure loads and settlements. Use different sand and a clay with different water content.

Try pushing a pencil into a sugar bowl; the resistance is initially very small but quickly increases; why is this?

17.9 FURTHER READING

The basic principles and equations for bearing capacity and settlement of foundations are covered in most standard textbooks on geotechnical engineering. There are specialist textbooks on foundation engineering, on shallow foundations and on piled foundations.

Not all of these deal simply and clearly with the general case in Section 17.3 and few include a clear explanation of the influences of groundwater and free water above ground level.

The basic bearing capacity equations can be checked by inserting values of the bearing capacity factors for $s_u = 0$ or $c' = \phi' = 0$ in which case the equations should reduce to Archimedes principle.

Chapter Eighteen

Retaining Walls

If an unsupported slope is not safe it can be supported by a retaining wall. Retaining walls also store water and exclude water from excavations.

18.1 TYPES OF RETAINING WALL

Retaining walls are made from mass concrete, steel sheet piles, concrete piles or concrete panels. They may be supported by shear stress on the base, earth pressures in front and by props and anchors.

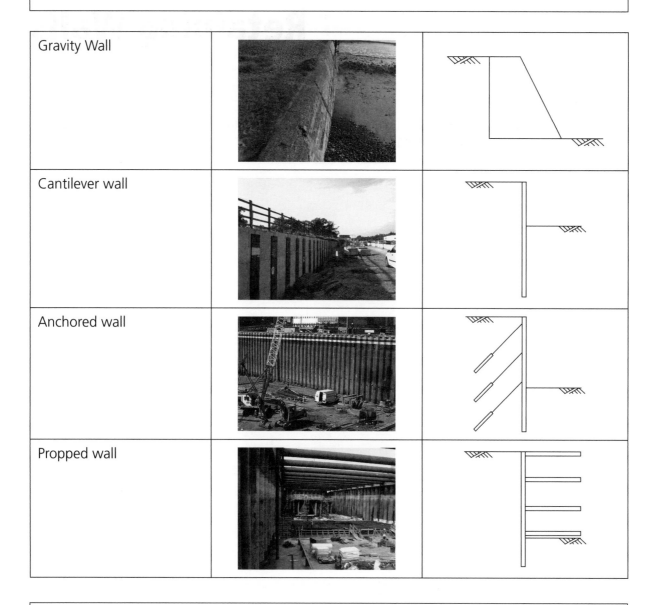

Gravity Wall		
Cantilever wall		
Anchored wall		
Propped wall		

Retaining wall design requires calculation of overall stability to prevent collapse due to slipping through the soil and calculation of forces in props and anchors and bending and shear stresses in the wall itself. Design may also require calculation of ground movements.

If the soil is drained effective stresses depend on pore pressures and water may leak below the wall into the excavation. Both require analyses of steady-state seepage.

18.2 EARTH PRESSURES ON RETAINING WALLS

The stresses that soil and water apply to a retaining wall depend on the soil strength, on whether the soil is drained or undrained and on whether the wall is moving towards or away from the soil.

Passive pressures σ_p act on the side towards which the wall is moving.

Active pressures σ_a act in soil on the side away from the movement of the wall.

If there is no movement the pressure σ_o is the earth pressure at rest (Section 10.5).

As the wall moves, passive pressures increase and active pressures decrease and both reach limiting states at which the soil strength is fully mobilised.

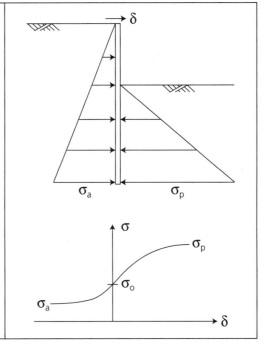

Horizontal pressures on a wall are total stresses and free water stresses; total stresses are the sum of effective stresses σ' and pore pressures u.

If the wall is rough there may be shear stresses between the wall and the soil.

The limiting active and passive pressures and shear stresses occur when the soil reaches its strength. If the soil is drained earth pressures depend on ϕ' and if the soil is undrained they depend on s_u.

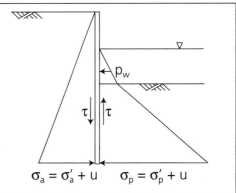

There are two routine methods for simple calculation of earth pressures.

Rankine earth pressures: the analysis considers equilibrium stress states that do not exceed the failure of the soil; this is a lower bound plasticity calculation (Section 15.3).

Coulomb wedge analyses: the analysis considers equilibrium of forces on a mechanism consisting of straight slip planes; this is a limit equilibrium calculation (Section 15.2).

18.3 RANKINE EARTH PRESSURES

These analyses are for simple cases of a smooth wall and level ground in front of and behind the wall. They calculate the variation of active and passive horizontal total and effective stress with depth. There are more complex solutions for rough walls and sloping ground.

	Passive pressures	Active pressures
Drained loading Total passive earth pressure σ_p from $\sigma_v' = \sigma_v - u$ $\sigma_p' = K_p' \sigma_v'$ $\sigma_p = \sigma_p' + u$ and similarly for σ_a'	 $K_p' = \dfrac{\sigma_p'}{\sigma_v'}$ $K_p' = \dfrac{1+\sin\phi'}{1-\sin\phi'} = \tan^2\left(45 + \tfrac{1}{2}\phi'\right)$	 $K_a' = \dfrac{\sigma_a'}{\sigma_v'}$ $K_a' = \dfrac{1-\sin\phi'}{1+\sin\phi'} = \tan^2\left(45 - \tfrac{1}{2}\phi'\right)$
Undrained loading	 $\sigma_p = \sigma_a + 2s_u$	 $\sigma_a = \sigma_v - 2s_u$
Tension cracks	Since σ cannot be negative a crack can form on the active side which can fill with water. The critical depth is when $\sigma_h = \gamma_w H_c$ and $\sigma_v = \gamma H_c$ and $H_c = \dfrac{2s_u}{\gamma - \gamma_w}$	

18.4 COULOMB WEDGE ANALYSES

These analyses satisfy equilibrium of the forces acting on slipping wedges. They determine the magnitudes of the horizontal active and passive forces but not their point of action.

To obtain a limit equilibrium solution the angle of the wedge must be varied to maximise the passive force and minimise the active force.

	Active pressures	Passive pressures
Drained loading $P_p = P'_p + U_p$ $N = N' + U_n$		
Undrained loading $T = s_u L$ L = length of slip plane		

18.5 OVERALL EQUILIBRIUM OF EMBEDDED WALLS

The forces arising from active and passive earth pressures, props and anchors, pore pressures and free water pressures must satisfy horizontal and moment equilibrium. For rough walls where there are shear stresses on the wall and inclined anchors these should strictly satisfy vertical equilibrium but this is not often considered.

Calculation of moments

Earth pressure forces and moments are calculated from Rankine earth pressures.

The example shows two soil layers on the active side. For drained loading $P = P' + U$.

Moments about the top of the wall. (It is convenient to divide the stresses in soil B into a rectangle and a triangle.)

Counter-clockwise moment $= \Sigma P_a z_a$

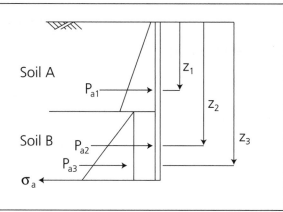

Propped and anchored walls

Equilibrium of moments at the prop level gives the depth d to satisfy

$\Sigma P_a z_a = \Sigma P_p z_p$

Resolving horizontally gives the prop or anchor force

$P_r = P_a - P_p$

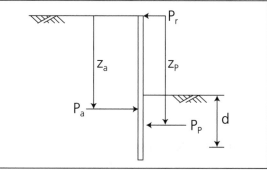

Cantilever walls

To satisfy both horizontal and moment equilibrium the wall must rotate about point R above the toe. The earth pressures below R can be replaced by an equivalent force at R.

Moments about R gives the depth d to satisfy

$P_a x_a = P_p x_p$

The design depth of the wall is 1.1d to 1.2d.

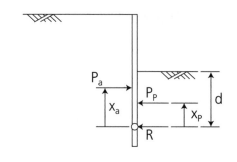

Bending moments and shear forces in embedded walls are found from structural analysis considering the wall as a beam with distributed loads.

18.6 OVERALL EQUILIBRIUM OF A GRAVITY WALL

The forces on a gravity wall must be in equilibrium. They must prevent overturning which is similar to toppling failure of a rock slope. The shear stresses on the base of the wall must prevent sliding.

Normal stresses at the base cannot be negative but may be zero. The normal stress must not exceed the bearing capacity of the ground (Section 17).

Undrained loading

For given soil parameters and retained height the design determines the width L.

At the limiting state the bearing capacity q_c (Section 17.3) is mobilised over an effective width L' and $Q = q_c L'$.

Moments about the center of the base $P_a x_a = Qe$ and the effective width $L' = L - 2e$.

Resolving forces vertically

$W = Q = q_c L'$

To prevent sliding on the base

$P_a = T = S_u L$

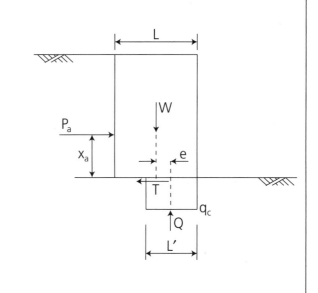

Drained loading

The soil stresses above are now effective stresses and total stresses are these plus pore pressures.

For steady-state seepage these usually may be approximated to triangular distributions.

The analyses for bearing capacity and sliding are similar to those for undrained loading.

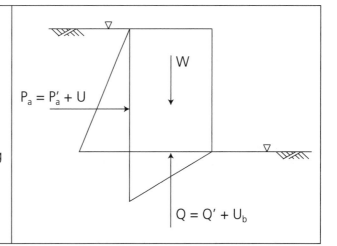

18.7 INFLUENCE OF WATER ON RETAINING WALLS

Free water applies a total stress to the front of the wall which generally improves stability. Pore pressures apply stresses to both sides of the wall. Pore pressures reduce the effective stress in the soil and its strength. Usually retaining walls include drainage to reduce the pore pressures in the soil.

Undrained soil

Pore pressures in the soil respond to maintain constant water content and constant strength and are unknown.

Free water in the excavation applies a total stress $p_w = \gamma_w h_w$ and a force

$$P_w = \tfrac{1}{2}\gamma_w h_w^2$$

There may also be free water pressures in tension cracks.

If the wall is impermeable P_w acts on the front of the wall and if the wall is porous it acts directly on the soil. In both cases the force on the soil is the same.

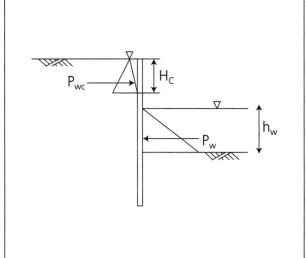

Drained soil

Pore pressures in the soil are governed by equipotentials found in a steady-state seepage flownet (Section 7.4).

On any equipotential $P = \dfrac{u}{\gamma_w} + z$

The change in potential between any two adjacent equipotentials is

$$\delta P = \frac{\Delta P}{N_d}$$

Pore pressures and effective stresses in the ground are different for permeable and impermeable walls because the flow nets are different.

Impermeable wall

Permeable wall

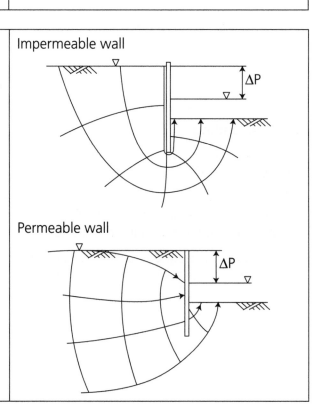

18.8 FURTHER READING

Warning: The simple analyses given above do not cover all the factors that have to be considered in design of retaining walls. There is a lot that can, and does, go wrong.

There are several specialist textbooks and some countries have codes and standards specifically on design and construction of retaining walls. There are several software programs for analysis of retaining walls. Many of these make different assumptions and simplifications and apply factors in different ways. They often produce different designs for the same height of retained soil.

Safe and economic design of retaining walls is one of the most challenging tasks in geotechnical engineering.

Chapter Nineteen

Unsaturated, Improved and Difficult Soils

The theories developed elsewhere in this book are for saturated and unstructured soils. Near ground level soil may be unsaturated especially in hot dry regions and near vegetation. Soils are often weakly bonded. Soils deposited from water often contain layers of coarser and finer grains.

Ground that is not sufficiently strong or stiff for the intended works can be improved or strengthened in situ or as fill or by loading. These processes and techniques are important for practice but detailed discussion is beyond the scope of this book. What follows in this section is a very brief summary as a guide to further reading.

19.1 COMPACTION

Natural soils are used to make embankments for road and rail and for flood protection and water storage dams. Soil is excavated, usually as saturated lumps and compacted by rolling or vibration in layers.

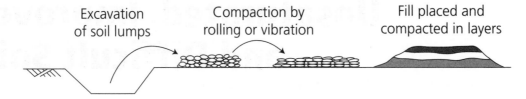

| Excavation of soil lumps | Compaction by rolling or vibration | Fill placed and compacted in layers |

The object of compaction is to manufacture fill that is sufficiently stiff and strong. Although the original soil was saturated the fill is not and it contains air, originally between the loose placed lumps that is not entirely removed by compaction.

Dry density ρ_d is a measure of the quantity of soil grains in a given volume of fill. The units of ρ_d are Mg/m³ and 1 Mg/m³ $\approx$ 10 kN/m³.

For a given compactive effort the dry density varies with water content as a compaction curve.

There is an optimum water content at which the dry density has a maximum.

There are relationships between dry density and water content for soil with varying degrees of saturation including $S_r = 100\%$ (Section 3.5).

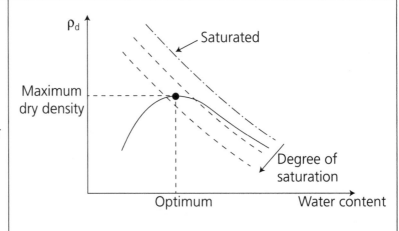

19.1 COMPACTION (CONT)

Soils are usually compacted at water contents close to optimum. In deserts they are compacted on the dry side because of the scarcity of water and in other places they are compacted on the wet side of optimum because it can be difficult to reduce the water content.

Note that here dry and wet sides of optimum is not the same as dry and wet sides of critical (Section 12.3).

Control and monitoring of compaction

Designers determine whether the required degree of compaction is achieved by a method specification or a performance specification.

1. A method specification determines the type and weight of roller, the thickness of each layer and the number of passes of the roller and these are checked during construction.

2. A performance specification determines the required minimum dry density or the minimum unconfined compressive strength and these are measured in situ and on samples during construction.

19.2 UNSATURATED SOIL

Soils can remain saturated for considerable distances above the water table (Section 7.2) and the basic theories in this book are for soils that are saturated with porewater pressures u_w. In an unsaturated soil air is present which has a pressure u_a. In this case the simple effective stress equation $\sigma' = \sigma - u$ no longer applies. Currently, there is no simple and robust effective stress theory for unsaturated soil.

The arrangements of air and water between the soil grains depend on the degree of saturation.

Air is continuous   Both water and air   Air is in bubbles
                       are continuous

Definitions

Net stress $= \sigma - u_a$

Matrix suction $= u_a - u_w$

Compacted soils are initially unsaturated while saturated soils can become unsaturated by evaporation in hot regions with low rainfall and by removal of water by vegetation.

If the water content is low and the air is continuous $u_a = 0$ and u_w is determined by the water meniscuses.

If the air is present as bubbles $u_a \neq 0$.

19.2 UNSATURATED SOIL (CONT)

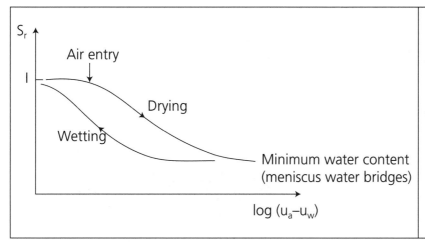

Matrix suction is related to degree of saturation but there are different relationships when the soil is drying and when the soil is wetting.

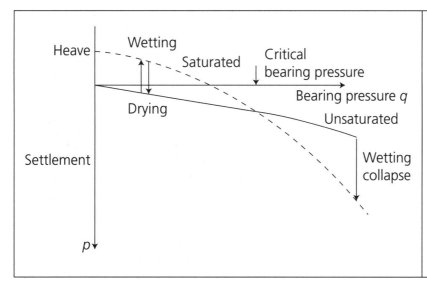

As soil dries from a saturated state it shrinks and foundations with relatively low bearing pressure settle.

As unsaturated soil wets towards a saturated state it swells and foundations with relatively low bearing pressure heave.

More heavily loaded foundations on unsaturated soil settle due to wetting collapse.

19.3 GROUND IMPROVEMENT

If the ground is too weak or too soft for the intended works and if deep foundations or piles (Section 17.1) are not feasible or economic, the ground may be improved in a variety of ways.

There are several commercial processes for ground improvement. These generally follow one of a few basic methods.

Reinforced fill

Metal or plastic strips are placed in coarse-grained fill and support structural front panels.

Stabilisation

Fine-grained fill is mixed with a cementing material such as cement or lime.

Reinforced in situ

Soft ground is reinforced in situ by installation of columns of stone or soil is mixed in situ with cement or lime.

19.4 GROUND IMPROVEMENT BY LOADING

Staged loading to prevent undrained failure

If rapid construction of an embankment would cause a bearing capacity failure (Chapter 17), the embankment is first constructed to a lower safe height and the ground allowed to gain strength by consolidation and reduction in water content.

During consolidation the water content and strength vary with time and depth in the ground but there is an increase in mean strength with time.

At a time t_1 the soil has gained sufficient strength to provide adequate bearing capacity for the additional fill. Tall embankments on weak soil can be built in several stages.

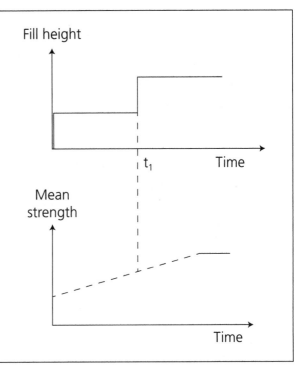

Pre-loading to reduce post-construction settlement

Fill is first placed to a height above the final design height and the ground consolidates and settles under this load.

At a time t_2 when consolidation is not complete but the settlement corresponds to the design final settlement the additional fill is removed.

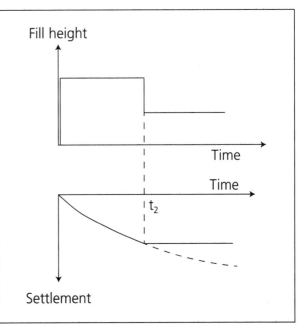

19.5 STRUCTURED SOILS

Structured soils have bedding or layering and they may be bonded (Section 3.7). The influence of structure can be investigated by comparing the behaviour of an undisturbed sample with that of a sample of the same soil that has been destructured. This can be done by mixing the soil at a large water content and recompressing it to the same state.

Layers of coarse- and fine-grained soil

The state boundary surface of a sedimented layered soil is larger than that of the same soil reconstituted because the fine grains do not occupy spaces between coarse grains.

On loading with a constant stress ratio τ'/σ' the reconstituted and sedimented soils yield at different states at B and C.

With continuing loading the layers remain and do not merge into one another. The reconstituted soil follows B–B and the sedimented soil follows C–C.

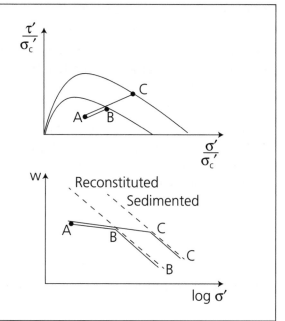

Bonding between grains

The state boundary surface of a bonded soil is larger than that of the same soil reconstituted because the bonding increases the yield stress.

On loading with a constant stress ratio τ'/σ' the reconstituted and bonded soils yield at different states at B and C.

With continuing straining the bonding deteriorates. The reconstituted soil follows B–B and the bonded soil follows C–C which approaches the reconstituted soil as the influence of bonding diminishes.

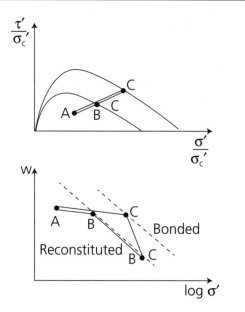

19.6 FURTHER READING

Few standard textbooks deal with unsaturated soils, structured soils and ground improvement in any detail. There are specialist books that deal with these topics. Some of the processes employed in ground improvement are the property of specialist contractors who have their own design methods.

Chapter Twenty

Principles of Geotechnical Design

The basic analyses in geotechnical engineering (Section 15) examine:

- Collapse of foundations, slopes and walls when soil strength is fully mobilised
- Movement of foundations when soil has a stiffness that corresponds to the strains in the ground
- Seepage and consolidation

Geotechnical design requires that structures have a margin of safety against catastrophic collapse and they do not move so much that their performance is compromised. These are quite separate requirements.

At the same time geotechnical structures should be reasonably economic; it is wasteful to take more land than required for a slope or to pour unnecessary concrete into the ground. Geotechnical structures, like all civil engineering works, must be safe, serviceable, economic, sustainable and buildable by the local work force.

20.1 WHAT IS FAILURE?

Failure has a general meaning. It includes both collapse (an ultimate limit state) and too much movement (a serviceability limit state) when the structure fails to do what it is intended to do but without totally collapsing.

As a foundation is loaded or a slope made deeper or steeper there will be movements. If the movements become very large the structure is said to have collapsed at its ultimate limit state.

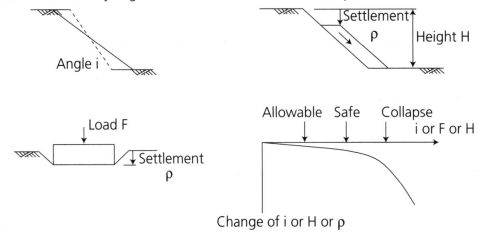

Collapse load

The collapse load is the calculated bearing capacity of a foundation or limiting height and angle of a slope. It may be the buckling load of a prop or the bending strength of a retaining wall; both would cause the wall to collapse if exceeded.

Safe load

The safe load is less than the collapse load. It ensures that the structure is not dangerous although deformations may be relatively large.

Allowable load

The allowable load is not more than the safe load and is usually considerably less. It places the design in a region of the load settlement curve where movements will be small.

In most cases geotechnical design is governed by the requirement to limit movements at working load (the serviceability limit state).

Normally, if movements are limited the ultimate limit state will be satisfied too: if it does not move much it will not collapse but the ultimate state should always be checked independently.

In some cases, such as a slope in open country the only requirement is to prevent collapse at the ultimate limit state.

20.2 USE OF FACTORS TO PREVENT FAILURE

Analyses for design of foundations, slopes and walls include factors. Their purpose is to

- Ensure safety
- Limit movements
- Allow for uncertainty

Factor of safety F_s

A safe load is found by applying a factor of safety to the calculated collapse load. The principal purpose is to allow for uncertainty and to ensure the structure is safe.

Load factor L_f

An allowable load is found by applying a load factor to a collapse load. The principal purpose is to limit movements. An allowable load may also be found using soil stiffness to calculate movements.

The purpose of a factor of safety is to ensure safety; this is different from the purpose of a load factor which is to limit movements. (In many books a load factor intended to limit movements is called a factor of safety but this is misleading.)

Warning: Safety factors and load factors are applied in different ways in different codes and standards.

Design using overall factors

In many codes and design processes a factor of safety or a load factor is applied once. It may be applied to a soil strength or to a load or to a stress. In the case of a foundation, it may be applied to a net or to a gross bearing capacity.

Design using partial factors

Some codes, instead of applying a single overall factor of safety or load factor, apply separate partial factors to loads, to soil parameters and to resistances. An example of this approach is the Eurocode family. Other national and local codes may apply factors in different combinations.

Warning: The same value of a factor of safety or a load factor can be applied in different ways and design solutions vary accordingly (Sections 20.5 and 20.6).

20.3 CHANGES WITH TIME

Normally construction of slopes, foundations and walls in fine-grained soil will be undrained and there will be excess pore pressures that may be positive or negative. As time passes after construction these dissipate until they become zero and the pore pressures become steady state.

Excavated slopes and walls (Section 16.2)

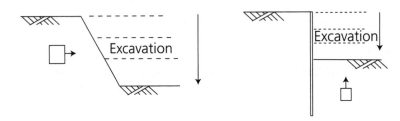

The arrows indicate the *changes* of normal stress; shear stress increases in every case.

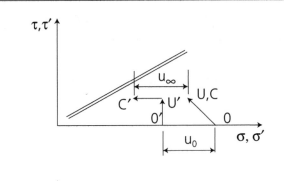

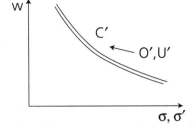

O'U' = undrained loading (assuming isotropic and elastic soil)

U'C' = consolidation

Steady-state pore pressure after consolidation u_∞ may or may not be the same as the initial steady-state pore pressure u_0.

If undrained construction involves reduction of mean total stress pore pressures generally reduce during construction. Then during consolidation pore pressures increase with time after construction, the stress path U'C' moves towards the failure line, the water content increases, the soil swells and becomes weaker; the structure becomes less safe.

20.3 CHANGES WITH TIME (CONT)

Foundations and filled walls (Section 17.2)

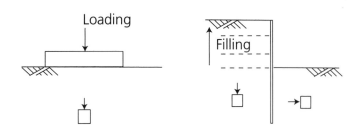

The arrows indicate the *changes* of normal stress; shear stress increases in every case.

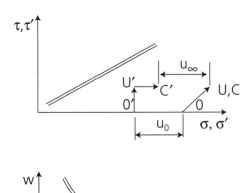

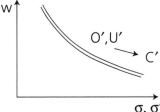

O'U' = undrained loading (assuming isotropic and elastic soil)

U'C' = consolidation

Steady-state pore pressure after consolidation u_∞ may or may not be the same as the initial steady-state pore pressure u_0.

If undrained construction involves increase of mean total stress pore pressures generally increase during construction. Then during consolidation pore pressures decrease with time after construction, the stress path U'C' moves away from the failure line, the water content reduces, the soil compresses and becomes stronger; the structure becomes safer.

20.4 SELECTION OF DESIGN PARAMETERS

Soil has several strengths (peak, critical state and residual—Section 11.1) and soil stiffness varies with strain (Section 10.6). For each unit in the geotechnical model (Section 4.1) there may be a range of values of each strength found from laboratory tests, in situ tests and by correlation with soil classifications. From these data geotechnical designers select characteristic or design values (Section 14.5) to which they apply overall or partial factors.

The engineer has to decide which strength (peak, critical state or residual) and which stiffness to adopt as the relevant characteristic or design value. These choices are governed by the requirements of the design and the factors that will be applied.

Design to prevent	Analysis	Design parameter	Typical factor
Collapse	Soil strength + factor of safety	Peak strength	$F_s = 1.25–1.5$
		Critical state strength	$F_s \approx 1.0$
Reactivation of existing slip planes	Soil strength + factor of safety	Residual strength	$F_s \approx 1.0$
Excessive ground movement	Soil strength + load factor	Peak strength	$L_f \approx 3$
Excessive ground movement	Stiffness	Stiffness at a strain $\approx 0.1\%$	No factor

Warning: Choice of characteristic or design values of soil parameters, and appropriate factors is not easy. Few standards and codes give clear advice.

20.5 DESIGN OF A SHALLOW FOUNDATION FOR UNDRAINED LOADING

Factors of safety or load factors can be applied to loads, or stresses or soil strengths in several different combinations giving different design loads for the same factor.

In the example the undrained collapse load F_c of a simple foundation is 640 kN and the design load F_d is calculated with a factor = 2 applied in different ways.

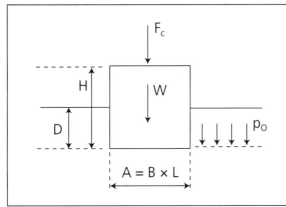

For vertical equilibrium at collapse when the soil strength is fully mobilised:

$$q_c - p_o = s_u N_c \quad \text{or} \quad F_c + W - \gamma DA = s_u N_c A$$

For a shallow square footing $N_c = 6$.

Dimensions of footing	Ground properties
$L = B = 2$ m and $A = 4$ m^2 $H = 2$ m and $D = 1$ m $W = 160$ kN	$\gamma = 20$ kNm^{-3} $p_o = 20$ kPa $s_u = 60$ kPa $N_c = 6$
$F_c = \gamma DA - W + s_u N_c A$ and $F_c = 1360$ kN	

Factor = 2 applied to		Design load F_d with factor
Strength	s_u	640 kN
Gross bearing pressure	q	600 kN
Net bearing pressure	$q_n = q - p_o$	640 kN
Applied load	F	680 kN

20.6 DESIGN OF A SHALLOW FOUNDATION FOR DRAINED LOADING

The analysis for undrained loading is applied to the same foundation for drained loading. The soil is saturated with the tater table at the ground surface.

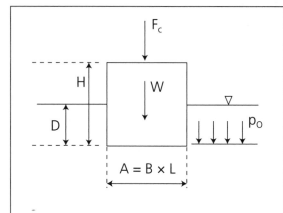

For vertical equilibrium at collapse when the soil strength is fully mobilised:

$$q'_c - p'_0 = \tfrac{1}{2}B(\gamma - \gamma_w)N_\gamma + (N_q - 1)p'_0$$

$$q'_c - p'_0 = (q_c - u) - (p_0 - u) = (q_c - p_0)$$

$$F_c + W - \gamma DA = A\left[\tfrac{1}{2}B(\gamma - \gamma_w)N_\gamma + (N_q - 1)p'_0\right]$$

where

$$N_q = \tan^2\left(45 + \tfrac{1}{2}\phi'\right)e^{\pi \tan\phi}$$

$$N_\gamma = 2(N_q - 1)\tan\phi'$$

Dimensions of footing	Ground properties
$L = B = 2$ m and $A = 4$ m^2	$\gamma = 20$ kNm^{-3}
$H = 2$ m and $D = 1$ m	$p_0 = 20$ kPa; $u = 10$ kPa and $p'_0 = 10$ kPa
$W = 160$ kN	For $\phi' = 30°$ $N_q = 16$ and $N_\gamma = 17$
	For factor $= 2$ $\phi' = 16°$ $N_q = 4$ and $N_\gamma = 4$

$$F_c = \gamma DA - W + A\left[\tfrac{1}{2}B(\gamma - \gamma_w)N_\gamma + (N_q - 1)p'_0\right] = 1200 \text{ kN}$$

Factor = 2 applied to		Design load F_d with factor
Strength	$\tan\phi'$	200 kN
Gross bearing pressure	q	520 kN
Net bearing pressure	$q_n = q - p_0$	560 kN
Applied load	F	600 kN

Notice that applying a factor to ϕ' reduces the design load considerably. This is because the bearing capacity factors are highly dependent on ϕ' (Section 17.4).

20.7 STANDARDS AND CODES

Most geotechnical engineering practice is governed by standards and codes of practice. These are different in different countries and regions. For example, in Europe there is Eurocode 7 and several related codes; in North America there are ASTM codes and standards. Some, such as the offshore oil and gas industry and road and rail owners have their own codes and standards.

Most standards and codes were written by committees drawn from across the geotechnical engineering profession many of whose members had commercial interests. As a consequence many are compromises.

The client who is commissioning and paying for the work should determine to which code and standard the work should be designed. Geotechnical engineers should ascertain the client's requirements at the start and they should be familiar with the requirements of the codes and standards which they have to satisfy. Nevertheless, all geotechnical investigations and designs should comply with the basic principles of soil mechanics and geotechnical engineering described in this book.

Keywords, Definitions and Index

Warning: In geotechnical engineering, as in all subjects, it is important to have clear definitions of terms and meanings of words. These should be consistent with their use in other engineering subjects. The following gives the definitions used in this book and the principal sections in which they are discussed.

Term	Definition	Section
1D compressibility	$m_v = \dfrac{\delta\varepsilon_v}{\delta\sigma_v}$ for loading with $\varepsilon_r = 0$.	10.1
1D modulus	$M' = \dfrac{\delta\sigma_v'}{\delta\varepsilon_v}$ for loading with $\varepsilon_r = 0$.	10.1
Active pressure	Horizontal stress on a wall when the wall moves away from the soil.	18.2, 18.3, 18.4
Allowable deformation	Movements that can be tolerated without loss of function; often called a serviceability limit state.	15.1, 20.1
Allowable load	The calculated load or geometry at which ground movements and deformations are allowable.	20.1
Bearing capacity	The *calculated* maximum load that can be applied to a foundation.	17.3
Bearing capacity factors	Factors for simple analyses to find a calculated bearing capacity for drained and undrained loading.	17.5
Bored pile	Hole bored in the ground filled with concrete and reinforced with steel cage.	17.7
Bulk modulus	$K' = \dfrac{\delta p'}{\delta\varepsilon_v}$ for isotropic loading.	10.1
Cam Clay	Elasto-plastic numerical model for soil.	13.2
Cantilever wall	Embedded wall without anchors or props.	18.1, 18.5

Term	Definition	Section
Characteristic value	The value of a parameter that characterises the overall behaviour of a relatively large volume of soil. (In some codes this has a more specific definition.)	14.5
Coarse-grained soil	Soil with less than about 30% fine grains; relatively large permeability; free draining.	3.3
Coefficient of consolidation	$c_v = \dfrac{k}{m_v \gamma_w}$	15.6
Coefficient of permeability	Constant in Darcy's law $V = ki$. k = seepage velocity with hydraulic gradient $i = 1$.	7.5
Cohesion	Strength that is independent of normal stress.	6.4
Collapse	Failure resulting in very large movements; often called the ultimate limit state.	15.1
Collapse load	The *calculated* load or geometry at which the soil strength is fully mobilised and movements are very large.	20.1
Compaction	Removal of air in fill by rolling, impact or vibration.	19.1
Compatibility	Deformation in which no gaps appear and no material is destroyed.	5.2
Consolidating	Pore water flow under the influence of excess pore pressure; pore pressure and rate of flow decay with time.	7.1
Consolidation	Change of volume due to drainage of excess pore pressure with constant total stress.	8.3
Constitutive equation	General relationship between changes of stress and changes of strain; includes stiffness and strength.	6.5
Coulomb wedge analysis	Analysis of earth pressure forces on retaining walls from limit equilibrium methods.	18.4
Coupled loading and drainage	Simultaneous loading and drainage; pore pressures and volumes both change.	8.3
Coupled shearing and volumetric effects	Change of shear stress causes volumetric strain and change of normal stress causes shear strain.	6.5, 9.2, 11.7, 13.2
CPT	Cone penetration test.	14.2

Term	Definition	Section
Creep	Strains continue at constant stress. Volumetric creep decays with time; shear creep may decay or accelerate.	6.9, 10.10, 12.5
Critical state line	Relationship between shear stress, effective normal stress and water content for soil deforming at its critical state.	11.2
Critical state strength	Strength for continuous uniform deformation.	11.2
Darcy's law	$V = ki$ for steady state seepage.	7.4
Deep foundation	Foundation in which the depth of the base of the foundation is not small compared to its width; shear stresses on the sides contribute to its capacity.	17.1
Description of soils	Contains: what the grains are; how they are packed together (loosely or densely); structure such as layering and bonding.	3.2 to 3.9
Design value	The value of a soil parameter for design. In some codes this is the value *after* a partial factor has been applied and in other cases it is the value *before* any factors have been applied.	14.5
Drainage path length	The largest distance travelled by a drop of water during consolidation.	15.6
Drained loading	Loading with constant pore pressures $\delta u = 0$. Pore pressures are known so analyses are in effective stresses.	8.3
Driven pile	Pre-formed pile driven into the ground.	17.7
Dry of optimum	Compaction at a water content less than the optimum water content.	19.1
Dry side of critical	States that have water contents less than the critical water content at the same effective stress. This is not the same as dry of optimum (Section 19.1).	12.3
Earth pressure	Total or effective stress in the ground and near a retaining wall. See: active pressure, passive pressure.	18.2

Term	Definition	Section
Effective stress	Effective stresses always have primes and $\sigma' = \sigma - u$ and $\tau' = \tau$. All changes of soil strength and soil volume are due to changes of effective stress.	8.1, 8.2
Elastic	A theory for material behaviour in which strains on loading are fully recovered on unloading.	6.6
Elastic–plastic	A theory in which material behaviour is perfectly elastic until yield and then is perfectly plastic.	6.7
Elasto-plastic	A theory in which elastic and plastic strains occur simultaneously.	6.7
Embankment	Fill placed to raise the ground level. (When the fill is stronger than the ground the stability analysis is like a shallow foundation.)	17.1
End bearing pile	Pile in which shear stress on the side makes a negligible contribution to its capacity.	17.1
Engineering geologist	Geotechnical professional usually with a first degree in geology and with skills in geological investigation and interpretation.	1.2, 1.3
Engineering rocks	Collections of grains or crystals that are strongly bonded. The rock mass contains discontinuities as joints or faults that control mass behaviour.	2.3, 3.1
Engineering soils	Collections of grains that are unbonded, or only very weakly bonded, and have water and possibly air in the pore spaces.	2.3, 3.1
Equilibrium	Forces resolved in any direction are zero: the body is not accelerating.	5.2
Equipotential	Line in the ground joining points of the same hydraulic potential.	7.4
Errors	The difference between a measured or calculated value and the true value. See random error, systematic error, gross error.	14.3
Estimated parameters	Values for several important soil parameters can be estimated from soil descriptions.	14.6
Excess pore pressure	In soil that is consolidating this is the difference between the current pore pressure and the long-term equilibrium pore pressure. It is denoted as $\bar{u}$.	7.1

Term	Definition	Section
Factor of safety	A factor applied in a collapse calculation to ensure the structure is safe; it may be applied to loads, stresses or soil parameters.	20.1, 20.2
Failure	A structure 'fails' when it does not meet the requirements for satisfactory performance. This does not necessarily imply catastrophic collapse.	15.1, 20.1
Fine-grained soil	Soil with more than about 40% fine grains; relatively low permeability and slow drainage.	3.3, 3.4
Flow rule	Relationship between plastic strains and yield stresses.	6.7, 13.2
Flownet	Graphical construction containing an orthogonal grid of flow lines and equipotentials with 'square' cells.	7.4
Foundation on elastic soil	Analyses of stress and deformation below foundations assuming the soil is elastic.	15.4
Free water	Water in rivers and flooded excavations; applies total stress to soil surfaces.	7.1
Friction	Strength that increases with normal stress.	6.3
Friction pile	Pile in which shear stress on the sides makes a major contribution to its capacity.	17.1
Geological classification	Classification based on mode of formation (e.g. igneous, sedimentary) and age (e.g. Jurassic, Carboniferous).	2.2
Geotechnical engineer	Geotechnical professional usually with a first degree in civil engineering and with skills in mathematics and mechanics.	1.2, 1.3
Geotechnical engineering	Construction on and in the ground using natural soils and rocks.	1.2
Geotechnical engineering professional	Engineer or geologist engaged in ground investigations and design and construction of civil engineering works in the ground.	1.2
Grading	Distribution of grain sizes; shown as a grading curve.	3.3
Grains	Individual particles described by: size, mineral composition, shape and surface roughness. Clay and silt are fine grains; sand and gravel are coarse grains.	3.4

Term	Definition	Section
Gravity wall	Retaining wall in which the resistance to sliding from shear stress between the base and the ground.	18.1, 18.6
Gross error	An error due to a major malfunction or mistake: result must be discarded.	14.3
Ground improvement	Strength and stiffness of the ground or fill improved by reinforcement, mixing with cement or loading.	19.3, 19.4
Ground investigations	Desk studies, walk-over survey, drilling, sampling, in situ and laboratory testing.	4.2, 4.4
Ground models	3D representation of the ground. Geological model based on geological descriptions; geotechnical model based on engineering behaviour.	4.1
Hardening	Relationship between the change of yield stress and plastic strain.	6.8, 13.2
Hydraulic gradient	$i = \dfrac{\delta P}{\delta s}$ along a flowline.	7.4
Hydraulic potential	$P = z + h_w =$ height of water in a standpipe above a datum.	7.4
In situ tests	Tests carried out in the ground.	14.2, 14.4
Infinite slope	A slope with a shallow slip plane parallel with the surface; characterised by a limiting angle.	16.3
Isotropic compression test	Laboratory test in which axial and radial stresses are equal.	9.1
Isotropic normal compression line	The state boundary for isotropic compression; separates possible from impossible states; yield curve for isotropic loading; slope C_c.	10.2
Laboratory tests	See: isotropic compression test, shear test, oedometer or 1D compression test, triaxial compression test.	14.2, 14.4
Limit equilibrium analysis	Analyses that satisfy equilibrium of a critical mechanism with the soil strength mobilised on slip planes.	15.2
Load factor	A factor applied to a collapse load calculation to ensure that the deformations of a structure or the ground are allowable.	20.2

Term	Definition	Section
Loading configurations	Isotropic; 1D, shear, triaxial, plane strain.	6.1
Material parameter	Soil parameter that depends only on the grains and is independent of state and structure.	14.1, 14.6
Maximum dry density	Largest dry density of compacted fill; this is achieved at the optimum water content.	19.1
Moderately conservative value	A value of a soil parameter somewhere between the worst credible and the representative values.	14.5
Mohr circle	Graphical construction to determine the shear and normal stress on any plane.	5.4
Non-linear stiffness	Soil stiffness that decays with strain.	10.6
Normally consolidated	The current state is on the normal compression line.	10.2
Numerical model	A set of equations relating changes of stress to changes of strain.	13.1
Oedometer or 1D compression test	Laboratory test in which $\varepsilon_r = 0$.	9.1
Optimum water content	Water content at which fill has a maximum dry density for a given compaction effort.	19.1
Overall equilibrium of retaining walls	Wall must satisfy equilibrium of horizontal forces and equilibrium of moments.	18.5
Overconsolidated	The current state is inside the normal compression line.	10.2
Overconsolidation ratio	$R_0 = \dfrac{\sigma'_m}{\sigma'} \geq 1$	10.2
Partial factor	Factors applied separately to loads and ground parameters to meet a design requirement.	20.1
Passive pressure	Horizontal stress on a wall when the wall moves towards the soil.	18.2, 18.3, 18.4
Peak strength	Maximum shear stress or stress ratio. Normally occurs at relatively small strains.	11.6, 11.8
Piles	See: end bearing pile, friction pile, driven pile, bored pile.	
Pile group	Several piles joined together at the top by a pile cap. Analyses may be the same as those for a deep foundation.	17.1

Term	Definition	Section
Plastic	A theory for material behaviour in which strains on loading are not recovered on unloading. Work done on loading is dissipated.	6.7
Plastic flow	Relationship between yield stresses and plastic strain.	6.7
Plasticity analyses (upper and lower bounds)	Analyses of the ground in which theorems of plasticity are applied to find states at which collapse can and cannot occur.	15.3
Plasticity index	Range of water content within which soil behaves as a plastic solid.	3.6
Poisson's ratio	$v' = -\dfrac{\delta\varepsilon_r}{\delta\varepsilon_a}$	10.1
Pore water	Pore water is in the pores of soil. It has a pore pressure u.	7.1
Pre-loading and staged loading	Ground strength and stiffness improved by loading and compression to reduce water content.	19.4
Propped or anchored wall	Stability of embedded wall improved by anchors or props.	18.1, 18,5
Random error	Small positive and negative departures from a true value; also called noise. Can be reduced by averaging.	14.3
Rankine earth pressures	Earth pressures on retaining walls determined from analyses of stresses.	18.3
Rate of loading and drainage	Times for construction and drainage determine whether soil is either drained or undrained.	8.4
Reports	Factual report; interpretive report; design report.	4.3
Representative value	Usually taken as the mean of a set of values or one that is close to the top of a distribution curve.	14.5
Residual strength	Strength of soils with laminar flow at large deformation when clay platelets have become aligned in a slip plane.	11.3

Term	Definition	Section
Retaining wall	Concrete or steel wall to support an otherwise unstable excavation or fill.	18.1
Safe load	The calculated load or geometry at which only a fraction of the soil strength is mobilised.	20.1
Saturated	Soil with pore spaces completely filled with water.	7.2
Seepage velocity	$V = \dfrac{\delta q}{\delta t.\delta A}$ = rate of flow through an area of soil δA.	7.4
Serviceability limit state	The design state for which movements are acceptable.	15.1
Settlement and heave	If the net bearing pressure is positive $(q > p_0)$ a foundation settles; if it is negative $(q < p_0)$ a foundation heaves.	17.3
Shallow foundation	Foundation in which the depth of the base is small compared to its width; shear stresses on the sides are ignored.	17.1
Shear modulus	$G' = \dfrac{\delta \tau'}{\delta \gamma}$	10.1
Shear test	Laboratory test in which shear stress is applied directly to the sample.	9.1
Shear vane test	In situ test to measure undrained strength.	14.2
Shear wave velocity	Velocity V_s of a shear wave which is related to the very small strain shear modulus G'_0.	10.7
Slope stability analyses	Analyses to find the height and angle of stable slopes.	16.3
Slope stability charts	Tables and charts for routine slope stability analyses.	16.3
Soil behaviour	During compression and swelling and during drained and undrained shearing.	9.2, 9.3, 9.4
Soil parameters	Numerical value describing strength, stiffness or permeability.	14.1
Soil strength	Limiting shear stress: depends on effective stress and water content and on strain.	11.1
SPT	Standard penetration test.	14.2

Term	Definition	Section
Specific volume	$v = \dfrac{V}{V_s}$ (where V_s = the volume of the grains) is a measure of the closeness of packing of grains from loose to dense.	3.7
State	Combination of current effective stress and water content: determines the basic behaviour of a soil.	12.1
State boundary	A boundary of all possible combinations of shear stress, effective normal stress and water content; includes the critical state line and the normal compression line.	12.3
State-dependent parameter	Soil parameter that depends on the grains and on the current state: may also depend on strain.	14.1, 14.6
State parameter	S_σ or S_w: measure the position of the current state with respect to the critical state.	12.2
Steady-state seepage	Porewater flow due to hydraulic gradients with no change with time: analyses find the rate of leakage and the pore pressures in the ground.	7.1, 15.5
Stiffness modulus	Relationship between changes of stress and changes of strain: may be a tangent or a secant.	5.3, 10.1
Strain	Shear strain γ describes distortion or change of shape: volumetric strain ε_v describes change of size.	5.3
Strength	Limiting resistance to a shear stress; may be tensile strength, compressive strength, bending strength and so forth.	5.3 6.3
Stress	Shear stress τ; normal stress σ.	5.3
Stress changes below foundations	Changes of total and effective stress in soil below loaded foundations.	17.2
Stress changes in slopes	Changes of total and effective stress in the ground behind stable and unstable slopes.	16.2
Stress in unsaturated soil	Net stress $= \sigma - u_a$ and matrix suction $= u_a - u_w$.	19.2
Stress–dilatancy	Relationships between stress ratio and rate of volume change provides model for peak and critical state strengths.	11.7

Term	Definition	Section
Structured soil	Soils which are in relatively thin layers with different gradings or in which the grains are bonded.	19.5
Suction	Same as negative pore pressure.	7.3
Surface processes	Natural processes of weathering, and deposition that manufacture engineering soils near Earth's surface.	2.4
Swelling line	Relationship between volume and stress for soil inside the normal compression line. Includes hysteresis and is often idealised to a single line with slope C_s.	10.2
Systematic error	Departure from a true value due to a known and quantified effect; can be reduced by corrections.	14.3
Time factor	$T_v = \dfrac{c_v t}{H^2}$ describes the rate of consolidation.	15.6
Total stress	Total stresses arise from external loads from structures or free water and act on the soil grains and the pore water.	8.1, 8.2
Triaxial compression test	Laboratory test in which a cylindrical sample is loaded by increasing axial stress with constant radial stress.	9.1
Ultimate limit state	A state of collapse.	15.1
Unconfined compressive strength	Strength of a cylindrical sample loaded axially with no radial stress.	11.5
Undrained loading	Loading with no drainage: pore pressures are not known so analyses are in total stresses.	8.3
Undrained strength	Strength of soil loaded with no drainage and at constant water content.	11.4
Unsaturated soil	Soil containing both water and a gas (usually air).	7.2, 19.2
Vertical cut and steep slope in undrained soil	An excavated slope in soil that is undrained has a maximum height and angle.	16.3
Very small strain stiffness	G'_0; shear modulus for strains <0.001%: found from measurement of shear wave velocity V_s.	10.6

Term	Definition	Section
Water pressures on retaining walls	Pore water and free water apply loads to walls; pore water influences effective earth pressures.	18.7
Water table	Surface in the ground joining all points where $u = 0$. If the water table is level the groundwater is hydrostatic and there is no seepage.	7.1
Wet of optimum	Compaction at a water content greater than the optimum water content.	19.1
Wet side of critical	States that have water contents larger than the critical water content at the same effective stress. This is not the same as wet of optimum (Section 19.1).	12.3
Worst credible value	It is unreasonable to assume a value for design worse than this.	14.5
Yield	A state at which behaviour changes from elastic to plastic or elasto-plastic.	6.6, 6.7, 13.2
Yield stress ratio	$R_y = \dfrac{\sigma'_y}{\sigma'} \geq 1$	10.2
Young's modulus	$E' = \dfrac{\delta q}{\delta \varepsilon_a}$	10.1